Can Science Make
of Life?

Sheila Jasanoff

———

CAN SCIENCE MAKE SENSE OF LIFE?

polity

First published in 2019 by Polity Press

Polity Press
65 Bridge Street
Cambridge CB2 1UR, UK

Polity Press
101 Station Landing
Suite 300
Medford, MA 02155, USA

ISBN-13: 978-1-5095-2270-5
ISBN-13: 978-1-5095-2271-2 (pb)

A catalogue record for this book is available from the British Library.

Library of Congress Cataloging-in-Publication Data
Names: Jasanoff, Sheila, author.
Title: Can science make sense of life? / Sheila Jasanoff.
Description: Cambridge, UK ; Medford, MA : Polity Press, 2018. | Includes
  bibliographical references and index.
Identifiers: LCCN 2018029799 (print) | LCCN 2018041646 (ebook) | ISBN
  9781509522743 (Epub) | ISBN 9781509522705 (hardback) | ISBN
  9781509522712 (pbk.)
Subjects: LCSH: Science--Philosophy. | Meaning (Philosophy) | Life.
Classification: LCC Q175.32.M43 (ebook) | LCC Q175.32.M43 J37 2018
  (print) | DDC 501--dc23
LC record available at https://lccn.loc.gov/2018029799

Typeset in 11 on 14 pt Sabon by
Servis Filmsetting Ltd, Stockport, Cheshire
Printed and bound in Great Britain by CPI Group (UK) Ltd, Croydon

For further information on Polity, visit our website: politybooks.com

Contents

Acknowledgments

Even a short book comes into the world carrying many unseen debts, and words alone are not sufficient to acknowledge them properly, though words must do. I must thank, first of all, Jonathan Skerrett at Polity for encouraging me to write this book in the first place and for his patience during the inevitable ups and downs of seeing the project through to completion. I am grateful to Sarah Dancy for her careful editing and Neil de Cort and the Polity production team for their invaluable backstage help. The book also represents years of research support from the US National Science Foundation, in particular under Award No. SES-1058762, as well as grants from the Greenwall Foundation and the Faraday Institute at the University of Cambridge. My thanks are due as well to Shana Ashar for the unfailing administrative support she provides in every aspect of my professional life.

Conceptually, the book benefited enormously from years of conversation with the students who have enlivened my course on bioethics, law, and the life sciences at the Harvard Kennedy School. I owe a special debt to my colleagues on the Harvard Embryonic Stem Cell

Acknowledgments

Oversight Committee, and to the scientists whose work we oversee, for sharpening my understanding of what is at stake in governing the moving frontiers of the life sciences and technologies. Words seem least adequate for thanking Stephen Hilgartner and J. Benjamin Hurlbut, two friends whose critical reading greatly improved the manuscript, but who are, more importantly, the interlocutors without whose answering enthusiasm my authorial imagination would long since have failed me. Hilton Simmet offered immeasurable help with the index, but more than that he opened my eyes to the possibilities of the book as metaphor and restored my sometimes flagging confidence in the value of the project. While the book owes much of its inspiration to my students, colleagues, and friends, any errors and inadequacies are of course entirely my own.

I am fortunate to be embedded in a family of writers who give meaning to a way of life that could strike others as isolating. That it rarely feels so to me is a tribute to the five people – Jay, Alan, Maya, Luba, and Nina – who constantly renew my wonder at what life is and what it is for.

Prologue

In February 1943, the Nobel Prize-winning quantum theorist Erwin Schrödinger delivered three lectures at Trinity College, Dublin on the advantages of borrowing terms from physics and chemistry to describe life. Published a year later, with the provocative title *What Is Life?*, the short book opened with a quote from Spinoza's *Ethics*: a free man's "wisdom is, to meditate not on death but on life." Schrödinger began his own meditation in a surprising place, with a disquisition on the smallness of atoms. He went on to speculate that life itself is organized at the molecular level in terms of regularities that explain both the variety and the relative stability of biological organisms. There must be some minimal ordering device, he suggested, capable of producing the living structures that we know and are. Two sets of chromosomes, the physicist observed, one from the mother and one from the father, "contain in some kind of code-script the entire pattern of the individual's future development and of its functioning in the mature state." How else to account for the infinite variety of life?

1

Prologue

The term "code-script" meant for Schrödinger that some directive force in the structure of the chromosomes in a fertilized egg must determine "whether the egg would develop, under suitable conditions, into a black cock or into a speckled hen, into a fly or a maize plant, a rhododendron, a beetle, a mouse or a woman" (Schrödinger 1967 [1944], 21). The miracle of biology, he suggested, is that some as yet unknown regulator inside each cell controls this wealth of living forms, as different in their diversity from the physical arrangements of periodic crystals as a Raphael tapestry is from a repeating wallpaper pattern. He likened this mechanism to a bureaucratic network operating through shared rules of the game: "Since we know the power this tiny central office has in the isolated cell, do they not resemble stations of local government dispersed through the body, communicating with each other with great ease, thanks to the code that is common to all of them?" (1967 [1944], 79).

Schrödinger's modest little book inspired a generation of young molecular biologists, most notably James Watson and Francis Crick, who cracked the code-script and so revealed the structure of life's "central office": deoxyribose nucleic acid, or DNA. But the moral implications of Schrödinger's essay lay elsewhere, in his conviction that the complex and abundant phenomenon we know as life could and would yield to material analysis at molecular levels. The code, together with the subtle laws that allow it to regulate all the stations of the body, accounts for the remarkable richness of life as we know it. In the book's concluding chapter, Schrödinger wondered about biology's ability to produce "order from order": "A single group of atoms existing only in one copy produces orderly events, marvellously tuned in with each other and with the environment according to most subtle laws" (1967 [1944], 79). But who would interpret those laws to deepen our understanding of life, and with what far-reaching consequences for the future of humanity? To those questions, Schrödinger offered no answer.

The twentieth century's great breakthroughs in the life sciences have made it increasingly more acceptable for biologists to claim ownership of the meaning of life. The origins and implications of that growing primacy deserve our attention. It is a story of arrogance in the literal, etymological sense (from Latin *ad* + *rogare*), a process of asking or claiming a terrain for oneself. Understanding how that happened and why it matters are the twin objectives of this book. The first, largely historical strand of my argument retraces the tangled pathways by which a particular way of interpreting life – that of the modern life sciences – acquired superiority over other, long-established discourses and modes of reflection. The second, more normative strand makes the case for restoring those more thoughtful ways of knowing, so that life does not devolve into just another object of conscious design, valued mainly for our ability to manipulate it, commodify it, and profit unequally from those acts of appropriation.

The beautiful simplicity of DNA's double helix provided, at first, its own compelling justification for biology's soaring status. The group of atoms Schrödinger spoke of contains, as we now know, paired bases that can be represented with just four letters: A, T, G, and C, standing, respectively, for adenine, thymine, guanine, and cytosine. If the matter of life could be broken down

3

into these letters, so regularly coupled A with T and G with C, then the idea of an entire book written in that parsimonious script proved to be almost irresistible. References to the book of nature, in which divine laws governing the workings of life are written down, had circulated in Western thought since medieval times. Unraveling the structure of DNA's double helix gave concreteness to the idea of nature's book. Revolutionary new knowledge brings new dreams of control, as the wily serpent saw when tempting Eve in the Garden of Eden. Biology in the post-DNA age offered similar temptations. Its central project of understanding was soon reimagined as one of authorship. Life scientists, as historians have noted (Kay 2000), quickly and enthusiastically adopted the metaphors of the book and the code, claiming the power not merely to read but to edit and eventually rewrite the book's contents.

The distinguished biologist Robert Sinsheimer was an early convert. On a rainy night in Pasadena in 1965, he gave a public lecture on "The Book of Life" comparing "the genetic information to the information in a book – like a book of recipes or a manual of flower arranging" (1994, 134). Almost imperceptibly, description merged into purposiveness, with the book of life recast as a how-to guide for humble makers and doers, such as cooks and florists whose creativity lies in recombining ingredients. Despite the promise of future applications, however, molecular biology continued to be celebrated as a science, a radically new form of knowing and sense-making. Nicholas Wade, longtime science writer for the *New York Times*, wrote a series of articles under the heading "Reading the Book of Life" (2000a) on the sequencing of the human genome. His first piece hailed the event as an "achievement that represents a pinnacle of human self-knowledge." Genomics, in this telling, was a path of enlightenment, a fitting launch for a new millennium. In another article by Wade (2000b), James Watson recollected his own singular role in that journey: "I would only once have the opportunity to let my scientific career encompass a path from the double helix to the three billion steps of the human genome." With Watson in the audience, President Bill Clinton acknowledged that extraordinary achievement on the day he announced the completion of a first draft of the mapped human genome. Alluding to the famously understated language of the 1953 Watson–Crick *Nature* article on the double helix, the president said to the scientist in three equally understated words, "Thank you, sir."

Less than twenty years later, Jennifer Doudna, a co-discoverer of the CRISPR-Cas 9 technique of gene editing, titled her account of the discovery *A Crack in Creation*. A portentous subtitle claimed for biology "the unthinkable power to control evolution" (Doudna and Sternberg 2017). Her book was not the first to link molecular biology with godlike power to make, or remake, humanity's destiny. In 1979, the journalist Horace Freeland Judson published a 686-page book named *The Eighth Day of Creation* that catapulted him to fame as a historian of contemporary science, while also making his title a byword for the biological revolution. Judson spent almost ten years interviewing most of the leaders in the field for his magisterial chronicle. Understandably, his focus was on science. Yet, stories emanating from the world's leading molecular biology labs were already hinting at unprecedented possibilities for manipulating living organisms. Indeed, Judson wrote in *Harper's Magazine* as early as 1975, "I think we are afraid of the plasticity of man" (1975a, 41). That fear was not misplaced. His big book followed by a year the first certified birth of a baby conceived outside the mother's womb, Louise Brown, born in England on July 25, 1978. The first successful cloning of a mammal from the cell of an adult animal, Dolly the sheep, born in Scotland on July 5, 1996, was less than two decades away, and five more years would lead to the mapping and sequencing of the human genome.

Accounts such as Judson's, and there are many of lesser note, display a scientific field busily scripting its own near-prophetic powers, a news media mesmerized by science-driven transformations in our understanding and expectations of life, and politicians hungry to take credit for advances that might win popular acclaim. Ever since the Luddites trashed the mechanical looms of the industrial revolution, reluctance to follow the lead of science and technology has been cast as misguided and retrograde (Juma 2016). But the rhetorics of hype and hope – and occasionally fear – that accompanied the fundamental biological discoveries of this era make it easy to lose sight of the complex social and cultural con-texts out of which the discoveries emerged, and which in turn shaped how the findings were turned to uses both bad and good. The capacity to rewrite the book of life proclaimed by the modern life sciences diverts attention from a history that includes darker chapters in which biology willingly partnered with state power: eugenic sterilization, racially motivated immigration laws, and Nazi experimentation, to name but the most salient few. Even Joseph Stalin's disastrous purges of Soviet genetics under the influence of Trofim Lysenko's anti-

Mendelian campaign can be read as a normal chapter in the accommodation between the promises of science and the aspirations of government, although Western commentary routinely dismissed Lysenkoism as an aberration, a deviant and one-sided appropriation of science by politics (Graham 2016).

The metaphor of the book performs in this connection its own imperial simplifications. Representing the human genome as *the* book of life, written in the plain four-letter code of DNA, implicitly claims for biologists a priestly role: as the sole authorized readers of that book, those most qualified to interpret its mysteries and draw out its lessons for the human future. But the genetic book of life sits in practice alongside numerous other volumes whose authors have also been occupied, for much longer stretches of time and across more diverse cultural spaces, in asking questions about the meaning and purposes of life in general and human life in particular. Some of those other books are also scientific, from fields such as ecology or evolutionary biology that are more inclined to view their subject matter as complex and systemic, hence not open to the genetic decoder's single master key. Other books in the ancient library of human thought approach their task of sense-making from perspectives stressing less what life *is* than what it is *for*. These are the books of law, religion, political theory, and moral philosophy, in which human societies have recorded since history began their ideas about what makes lives good or worth living – and, more specifically, what makes a life human and what is special about the condition of being human.

The power of the book metaphor, moreover, resides within a theological tradition that belongs to the peoples "of the Book," or the Bible. That association draws attention away from other ancient meditations on the meaning of life and the connections between its material and spiritual dimensions that are not as centrally mediated by books or codes. We may think here of the well-known episode in the Indian Upanishads where the teacher Uddalaka instructs his book-learned son Shvetaketu in the relations between an individual life and the absolute or supreme reality of existence. The son has returned, proud of his accomplishments in having studied the Vedas, the Hindu religious texts, when the father, through a series of examples, shows him that there is an essence or unity of being that is not the same as its particular manifestations. Most famously, the father asks the son to bring him a fruit of the *nyagrodha* tree and to break it open to see what is inside. The son sees a tiny seed and the father asks him to break that seed and say what it contains. The son does so and sees nothing. The father then tells him that immaterial essence, that visible nothingness, is in truth the essence of the tree and of all material, living selves. In the widely cited Sanskrit text, Uddalaka informs his son, "*Tat tvam asi,*" or "That art thou."

The point here is neither to put religious doctrines in competition with scientific theories nor to advocate for any particular dominant relationship between biology and religion or philosophy. It is far more to observe that descriptions of life have many origins and purposes, not all of them connected to unraveling or controlling the physical processes of being. Indeed, one distinction that has preoccupied Western philosophers from Aristotle through to recent and contemporary figures such as Hannah Arendt, Michel Foucault, and Giorgio

Agamben is precisely the difference between bare life (Greek *zoe*), the natural, physical life of the body, and the good or active life (Greek *bios*) that exists beyond the body, usually in relation to a community, a life committed to understanding and remaking its own condition. For students of social, political, and ethical life, meaning cannot be found in the bare essence of what makes us tick as atomized biological agents. *That* life is not human in some basic sense. To begin to examine the human condition is to note, with Arendt (1958, 22), that "[n]o human life, not even the life of the hermit in nature's wilderness, is possible without a world which directly or indirectly testifies to the presence of other human beings." Meaning, the answer to questions about life's purposes, germinates in that very connectedness.

This small book aims to correct, in a sense, the elegant but over-simplified optics of Schrödinger's physics-eye vision of life as code-script. Instead of asking "What is life?" *tout court*, my purpose is to show that this question cannot easily be disentangled from the linked and inseparable question, "What is life for?" Repeatedly over the last few decades of scientific development, human societies have confronted new frontiers of meaning as it becomes possible to arrange and rearrange the fundamental units of living matter in new ways. Where does life, as we care for it, begin? Where does that life end? How is one form of life, for example the human, related to other forms, including those of close biological similarity that do not show capacities such as language that we take to be definitive of human-ness? Linked to these morally charged questions are issues of social authority and responsibility. Whose opinion counts and whose does not in addressing these fundamental concerns?

Prologue

Does science have any special voice in defining human progress, and if so why? Who decides when answers are contested? Put another way, which interpreters of life's meaning are entitled to the highest authority when it is not clear whether an issue properly belongs to law or to science, to politics or to expert judgment, to shared social commitments or to private religious belief?

Unsurprisingly, none of these questions has proved amenable to easy answers and so the underlying issues remain very much alive. To advance our thinking, it is time to take stock of the multiple de facto authority claims and counterclaims that have sprung up around the practices of biology and biotechnology in the post-genetic era. In keeping with this book's title question, my focus will be on the role of science in settling (or claiming to settle) the ethical, legal, and social dilemmas that swirl around definitions of life. This science does by eliding the differences between natural and social life, and hence between what life is and what it should be for. We turn first to the emergence of biology not merely as a promising area of inquiry into the nature of life, but as a force that acquired superior cultural authority to determine the scope and limits of its own advancement. We look next at several areas, most notably reproductive and synthetic biology, where struggles for authority are currently taking place between biology, biotechnology, and other social institutions such as law and ethics that also have a stake in defining life's purposes. We conclude with observations on the institutional changes that are needed to address the unresolved tensions between the *is* and the *ought* of human existence at a time when biology is arrogating to itself nothing less than the power to control the evolutionary future of humanity.

Prologue

Developments in the life sciences and technologies, this book argues, are altering collective visions of desirable futures attainable through science and technology, or what we might call sociotechnical imaginaries (Jasanoff and Kim 2015), in contemporary societies. These changes run deep enough to affect constitutional understandings of who we are as human subjects and how we wish to be governed, not just as citizens of nation states but as living beings with the capacity to reflect on the value of our own existence and the meaning of our relations to nature and our earthly environment. As yet, the contours of those new understandings of science, technology, and society can be discerned only in dimmest outline. Bringing those nascent ideas into sharper focus, so that biology takes its rightful place within and not above society, is the hope and the aim of this book.

Figure 1.1   Where Do We Come From / What Are We / Where Are We Going (D'où Venons Nous / Que Sommes Nous / Où Allons Nous)

Source: Paul Gauguin, 1897

1

A New Lens on Life

In 1897, Paul Gauguin, in failing health and overcome with grief at the sudden loss of his favorite daughter, began painting a work he imagined as his last testament (Richardson 2009). Five feet high and twelve feet long, as wide as the inside of his Tahitian hut, the monumental painting laid out in a brilliant, dreamlike sequence many elements from the artist's personal symbolic repertoire: a sleeping child, mysterious figures whispering in the shadows, intertwined branches and tree trunks, an androgynous figure in the middle reaching up to pluck some fruit, scattered birds and animals, a stone idol with upraised hands, and an old woman marking the end of the allegorical life cycle. The colors are unearthly. Yellow-orange bodies glow against a background of acid blues, indigo, and deep green. Matte gold corners hint at a solid backing from which the tapestry-like composition seems to be peeling away. In the top left, Gauguin wrote the inscription that gives his masterpiece its name: *Where Do We Come From / What Are We / Where Are We Going.*[1]

On its face, this most iconic of Gauguin's Tahitian

paintings is a culmination of his long fascination with local mythology and its cultural milieu. The central figure, in particular, is a *mahu*, a man-woman or third-gender person, recognized and given special status in the cultures of the Pacific islands, though abhorred by colonizing Christian missionaries whose cultural common sense left no space for transgender persons (Geertz 1975). But the words that give the work its meaning seem to have sprung unbidden from the abandoned reservoir of Gauguin's early Catholic education in France. As a schoolboy, the future artist was taught by the eloquent and charismatic Bishop of Orleans, Félix-Antoine-Philibert Dupanloup, whose catechism for young students included questions about human origins and purposes: Where does humanity come from? Where is it going to? How does humanity proceed (Gauguin and Russell 2016)? In painting the arc of life at what he felt to be his own life's end, Gauguin reverted to these age-old questions about human existence that have preoccupied religious and philosophical minds since the beginnings of recorded thought.

Gauguin's last years in Tahiti, from 1895 to 1903, coincided with years of ferment in other worlds, the worlds of science, commerce, and industry from which the artist, sick and embittered, had fled. There, the nature of life was also the center of attention, but from entirely different angles and with very different consequences. Those were revolutionary years for modern biology, riding high on momentum built throughout the nineteenth century. The discoveries of that period set the stage, in turn, for the advances in the second half of the twentieth century that raised biology to a pedestal among the sciences previously reserved for atomic physics. Most important for our understanding of what life is and what it means, biology in those years came indoors, from the unruliness of the field to the systematic study of evolution, and ultimately into the sanitized orderliness of the lab, where new technologies increased our sense of mastery over life's processes. In turn, the manipulation of the matter of life in labs opened the door to commercial exploitation in medicine and agriculture and more ambitious plans for improving on nature's handiwork. Those shifts, the subject of this chapter, provide essential groundwork for understanding how biology positioned itself as the prime custodian of the meaning and purpose of human life and its place in the wider scheme of life on Earth.

Advances in the biological sciences fundamentally reshaped our thinking about the questions that perplexed Gauguin and his spiritual teacher: where life begins and ends; what is at stake in belonging to a species, kinship group, or family; what counts as normal or abnormal, healthy or diseased, and changeable or fixed in the natural order of things. The material descriptions of life offered by modern biology gradually took on prescriptive force, as if they were the foundation on which we should build our conceptions of good human futures, and as if those visions in turn should guide our technological interventions. Powerful new techniques for designing and redesigning life came to be seen as answers to old, value-laden questions, such as what counts as a well-lived life and who should be responsible for safeguarding lives on this planet. Science does not explicitly claim to offer full-blown answers to any of these questions, especially about the right ways to ensure life's protection and flourishing. Yet, as outlined in this chapter, biology and biotechnology have continually proclaimed themselves as humanity's most compelling instruments for making sense of life – those with the greatest power to answer the eternal questions posed by Gauguin's Tahitian masterpiece.

Origin stories: the evolution of life

From 1831 to 1836, long before Gauguin painted his enigmatic reflection on life and death, Charles Darwin undertook his famous voyage on the HMS *Beagle* to uncover in his own way one of life's basic mysteries: where do we come from? An avid beetle collector and botanist in his college years in Cambridge, Darwin acquired a passion for geology and the interpretation of strata well before setting sail on the *Beagle*. In the Galapagos archipelago, he was drawn to considering how diversity arose among living things, most famously in the finches he collected that are now named after him (Sulloway 1982). Discovery took root then; fame and adulation followed much later. Trained in theology as well as in natural history, and acutely sensitive to possible accusations of error, Darwin waited twenty years before going public in 1859 with his revolutionary work, *On the Origin of Species*.[2] Despite the outcry it provoked (and continues to provoke) in science, religion, and public culture (Wilson 2017), Darwin's claim that humans and other forms of life evolved through natural selection and adaptation proved hugely influential. Sigmund Freud, lecturing on the principles of psychoanalysis some sixty years later, called the theory

16

of evolution the "second discontinuity," on a par with the first discontinuity of the Copernican Revolution, which decentered Earth from its anchoring place in the solar system. Darwin's research, Freud wrote, had "robbed man of his apparent superiority under special creation, and rebuked him with his descent from the animal kingdom, and his ineradicable animal nature" (1920, 247). Evolution, in other words, was one of those rare breaks with past beliefs, a true scientific revolution.

The skeptical habits of thought that allowed Darwin to question foundational presuppositions about the biological origins of life did not extend to his theories about human cultures. Yet, here too an implicit commitment to lawlike progression could be detected. The Victorian moralist observed foreign human forms and practices with a dyspeptic eye from the secure perch of his own elevated position in an enlightened society. His adventures included a ten-day stop in Tahiti in November 1835, where, unlike Gauguin, he found the women "far inferior in every respect to the men" (1860, 430). He also commended Christian missionaries for having abolished "human sacrifices, and the power of an idolatrous priesthood," while reducing "dishonesty, intemperance, and licentiousness" in the indigenous populations (1860, 440). This kind of talk presupposes a kind of universalism in the dynamics of social progress. Theorists such as Herbert Spencer soon picked up on this thread, and "social Darwinism" emerged as a popular framework in accounting for progress.

Like any other transformative idea, Darwin's theory of evolution itself had a longer history and was carried out amidst other scientific efforts that help explain its hold on the modern imagination. Evolution was already in the air as an explanation for the complexity of life forms, in particular through the work of the French natural historian, botanist, and taxonomist Jean-Baptiste Lamarck in the early nineteenth century. Lamarck is now remembered largely for his discredited theory that acquired characteristics can be inherited. Even for those who held different views about the mechanics of evolution, however, he provided inspiration that complexity and diversity among organisms were not simply matters of chance or divine will. The development of life was governed by laws, and these biological rules of inheritance could be systematically studied and deciphered. Darwin was just one of the figures, if possibly the most renowned, who took up and carried forward that invitation to decode nature's laws, including the origins of species, through scientific scrutiny.

If Lamarck and later explorers like Darwin were preoccupied with variation among species across time and space, other pioneering naturalists of that period were more interested in how species pass on their characteristics through generations of offspring. A dozen years younger than Darwin, but with an active life more or less coincident with his, Gregor Johann Mendel, an Augustinian monk in St. Thomas's Abbey in the Moravian city of Brno, began studying the effects of crossbreeding pea plants in his monastery's small experimental garden. Encouraged by his teachers and colleagues, Mendel observed what happened when common edible peas carrying one set of distinctive traits – such as for plant height, flower color, or seed shape – were crossed with plants having the contrasting trait. After years of research, he discovered that traits which disappeared in the second generation reappeared in the third, in a proportion of roughly three to one. That finding led Mendel to his breakthrough conclusion: inheritable traits are transmitted through discrete "factors," which we now think of as genes, one "dominant" and the other "recessive"; plants exhibit the recessive trait, such as the green color in pea seeds, only if they contain two copies of the corresponding genetic variant or allele. Reproductive biology, in this respect, is destiny. Mendel's heritability studies boosted the view that laws of nature, not nurture, define important aspects of the kinds of beings we are, physically and perhaps even psychologically and intellectually. Our biological lineage marks us indelibly.

From field to lab: life redescribed

Today, Mendel's findings are foundational to the science of genetics, but the transformation of life, especially human life, into an object of experiment and explanation required significant displacements in the practices and objectives of biological research. Mendel himself achieved only posthumous fame. Unlike Darwin, who was born into a prosperous, well-connected family with the means to finance his voyage of self-discovery, Mendel came from modest circumstances and belonged to no influential scientific or social networks. His results were published in 1866 but, like his pea plants' recessive traits, both the research and its implications lay dormant for decades. It was not until 1900 that three European botanists working independently in three countries – Hugo de Vries in the Netherlands, Carl Correns in Germany, and Erich von Tschermak in Austria – rediscovered Mendel's papers and his laws of inheritance. Around the same time, Thomas Boveri in Germany and Walter Sutton in America established that chromosomes, threadlike structures within cell nuclei, contain the mechanism of heredity. Crossing continents, the Boveri–Sutton theory gained conclusive support from Thomas Hunt Morgan's experiments with the fruit fly at Columbia University in New York. This work established that chromosomes carry the genes, Mendel's so-called factors of heredity, at specific sites or loci. Parent organisms pass on one set of genetic variants, or alleles, to each offspring, accounting for the laws of heredity that Mendel had observed a half-century earlier.

Morgan's work, buttressed by his colleague Theodosius Dobzhansky's demonstration that natural selection takes place through gene mutations, cemented biology's already powerful hold on explaining life's origins. Importantly, too, it replaced the island, the countryside, and the botanical garden – the favored field sites of nineteenth-century natural history – with a new, more exciting place of discovery: the university research laboratory. Of course, the field to this day beckons the curious scientist who wishes to observe nature in the rough, but, as the historian Robert Kohler writes, "The field biologists who came of age in the early 1900s were the first who could not operate exclusively by their own rules on their own cultural ground. They lived in a world of laboratories, in which they felt bound to use lab methods and understood that their own practices and achievements would be judged by lab standards" (2002, 4). Young scientists found the lab world increasingly seductive. James D. Watson, co-discoverer of the molecular structure of DNA, was an avid bird-watcher in his youth, but tellingly he chose genetics as his field of graduate study in 1947.[3] From the turn of the century, the lab acquired near-hegemonic power to address the second of Gauguin's existential mysteries: what are we?

Lab practices changed both what one saw about life and how one saw it. Lamarck, Darwin, and Mendel based their theories on organisms they found or bred, over long periods, with eyes well trained to observe and classify visible variations. Lamarck studied plants and minerals for more than a decade before he published his compendium on French flora. The undergraduate Darwin compulsively chased down beetle varieties, and Darwin the explorer took more than twenty years to sort through his diaries for possible kinks in his theory of evolution (Richards 1983). Mendel spent eight years crossbreeding his pea plants before arriving at results he considered publishable. Training undoubtedly helped each of these pioneers, but the variations each saw in living things could be seen in principle by anyone taught what to look for. Finches might have bigger or smaller beaks, and peas could be smooth or wrinkled, and yellow or green in color; but a layperson could count how many of each kind existed in a place once shown which differences mattered and how to spot them in real life.

By the late nineteenth century, however, scientists concerned with life's propagation were in a position to extend the power of human eyes with instruments that could look inside whole organisms at structures invisible to the naked eye. Not only did these instruments reveal previously unseen things; they also altered what the historian Evelyn Fox Keller (2002, 211) has called "the biological gaze" and imposed their own disciplines of sense-making. The objects discovered in the lab were interpreted by experts trained in the use of specific techniques. The microscope, amplified by chemical techniques of dyeing organic matter, offered early glimpses into the internal structures of cells, and chromosomes (or "colored bodies") were identified and named years before their functions in heredity were well understood. Better lenses enabled "greater magnification and resolution, improved preparation of microscopic specimens, and, above all, renewed confidence in the veridicality of the basic instrument" (2002, 213). The arrival of cinematography allowed biologists not merely to see static patterns, but to watch the processes of life unfolding at a molecular level (Landecker 2013). Later still, x-ray crystallography made it possible to study complex molecules to get a clearer picture of their chemical structure. In John Randall's laboratory at King's College London, two young researchers, the New Zealander Maurice Wilkins and the Englishwoman Rosalind Franklin, used this technology to establish the helical structure of DNA and the placement of the phosphate units composing the backbone of the helix.

James Watson and Francis Crick completed the long and absorbing process of molecular discovery at the University of Cambridge. They derived crucial insights from the images made by Franklin, which the Cambridge pair, themselves strangers to crystallography, had to learn to read. Two-thirds of a century later, the resulting description of the structure of DNA is the stuff of every biology textbook: the twisted double helix of sugar-phosphate strands, connected ladderlike by four bases in matched pairs, adenine (A) to thymine (T) and guanine

(G) to cytosine (C). The four letters ATGC, abbreviating the names of the four bases, gave us an alphabet for "writing" DNA. More generally, they gave substance to the metaphor of life as a book, written in a code that had been cracked, and hence could be read, and, as we will see in later chapters, even edited by human ingenuity. Solving a chemical puzzle, however, was not the same as mapping the future ramifications of this knowledge. It was only a key for unlocking the door to an uncharted future. The 1953 article offered the barest hints of what was to come. The authors concluded with a famously understated sentence: "It has not escaped our notice that the specific pairing we have postulated immediately suggests a possible copying mechanism for the genetic material" (Watson and Crick 1953, 737). That implied mechanism for cell replication would eventually provide the basis for intentional interventions into the development of all living organisms, including even humans.

Twenty more years passed in the timeline of discovery before science acquired the tools to use DNA's copying mechanism with active intent, to improve desirable traits in plants and animals, to create entirely new entities and, eventually, to repair perceived malfunctions in nature's work. This was the period when discovery began to move from science to industry, although the lab remained crucial for deciphering the properties of living organisms, and even designing new ones. Capital entered the picture as medical and agricultural enterprises began eyeing the huge commercial potential from opening up the processes of life to targeted reengineering. Subtly but surely, the word *prospect* took on new meaning in relation to life. Once, a person's prospects referred mainly to the range of opportunities that one was born with and hence could hope for in life, from infancy to death. Today, biological prospecting refers to conscious acts of mining for profit the material and informational content of the genomes of all living things – in effect, supplementing the knowledge of what we are with imaginations of what else we might be.

Privatizing progress

The story told in this chapter is not merely about the rise of biology as the ultimate source of wisdom on where we came from, who we are, and where we are going; it is also a story of science uniting with capital to redefine our imaginations of progress. The global wars of the twentieth century both remade the map of the world and rewrote the political economy of nations. Older industrial states deprived of their colonial empires had to scramble for new sources of wealth to secure their citizens' welfare and prosperity. In an era of free trade and increasing competition, science and technology presented themselves as broad highways of opportunity, as reflected in Vannevar Bush's (1945) metaphor of science's "endless frontier." Promising jobs and growth along with cures for hunger and disease, biotechnology offered especially attractive prospects. European and North American governments eagerly seized on genetic engineering as a policy priority, recognizing all the while that they would have to balance the prospect of economic and social gains against the risks of misuse, whether through accidental releases of dangerous organisms or intentional acts of biological warfare. By the mid-1970s, recombinant DNA surfaced on the US regulatory agenda (Wright 1994), and over the next ten years both America and Europe debated approaches to managing a technology that combined the precision of molecular manipulation with the complexities and uncertainties of ecology, organismic biology, and human behavior.

The early years of commercial biotechnology coincided in the United States with growing opposition to perceived governmental overreach. The great age of expanding federal social programs and the rise of expert agencies that began with the New Deal in the 1930s ended decisively in 1980 with the election of Ronald Reagan. The new president's famously sunny persona masked a ruthless determination to roll back decades of health, safety, economic, and environmental regulation. This was a remarkably hospitable moment for launching a new industry armed with cutting-edge techniques and an almost limitless supply of imaginative applications. Genetic engineering promised a cornucopia of improvements in life and we know it now: pest-resistant and pesticide-tolerant plants that would repel infection and survive weed control treatments; farm animals that would produce more protein and less fat; effective, diagnostic, and therapeutic tools for fighting intransigent diseases; and, on the distant horizon, a hint of rolling back old age, even a hope of regeneration.

The nascent industry gained much mileage from the "bio" of biotechnology. Unlike the fearsome atom, with its explosive power to destroy all planetary life, and unlike toxic chemicals that were killing birds and silencing spring itself (Carson 1962), biotechnology posed as the natural alternative to the nasty, dirty technologies of the industrial past. Gene splicing and transfer only made more efficient and predictable the processes of mutation and adaptation that had been known since Mendel's time. Besides, so industry argued, people had been breeding plants and organisms for thousands upon thousands of years, for as long as human societies had practiced agriculture. The toolkit of molecular biology only enabled them to do the same things faster, in a more targeted fashion, with less time wasted in trial and error. An era of geneticization or genetic exceptionalism dawned (Lippman 1991), in which information contained in genes was deemed to have special power to predict and eventually cure inherited ills. Inventors and their sponsors, both public and private, saw all kinds of opportunities to rethink problems afflicting human societies in terms that invited genetic solutions, particularly hunger and disease but also, less obviously, issues such as aging and even urban violence (Allen 1999).

As relations among science, states, markets, and societies adjusted to make room for genetic knowledge and biotechnology, new questions of responsibility began looming. In democratic societies, it is the people's prerogative to articulate shared imaginaries of progress (Taylor 2004) and to elect governments to help them achieve their visions. In theory, the people decide whether their rulers are sufficiently responsive to actual public priorities and are following policies that correspond to public preferences. Now a powerful new axis emerged between science and industry with no such ties of political accountability to the people, even though visions of good human futures remained deeply contested. Private capital jumped in to support lines of research that publics as a whole had neither debated nor deemed desirable. Yet, as in the race to sequence the human genome that began between the National Institutes of Health (NIH) and Celera Genomics, the breakaway private company formed by one-time NIH employee J. Craig Venter, there was little discussion of what it means to turn science into a playground for private wealth. In the absence of such debate, biological research continued to propagate a myth of autonomy and capacity for self-governance. Almost imperceptibly, the notion gained ground that biology and biotechnology are uniquely equipped to define the proper directions and the ends of life – in short, to declare what constitutes human betterment. For life scientists and their enthusiastic promoters, the arc of the technologically possible, often coincident with the promise of financial gain, increasingly began to define the boundaries of the morally permissible.

To market, to market: life as commodity

Some of the most important steps in biology's seemingly inexorable march toward *biotechnology* took place on the American West Coast in the 1970s. Those advances carried new answers to Gauguin's third question: where are we going? At Stanford University, Paul Berg, co-winner of the 1980 Nobel Prize in Chemistry, found a way to splice DNA from a bacterium into the DNA of a tumor-producing virus, showing for the first time that it was possible to recombine DNA from two species. Herbert Boyer of the University of California at San Francisco and Stanley Cohen, also at Stanford, pooled their research skills to cut rings of bacterial DNA known as plasmids at specific loci using so-called restriction enzymes. This technique produced openings with "sticky ends" into which bits of foreign DNA could be inserted to rejoin the segments of the cut circle. The resulting recombinant-DNA (rDNA) molecule enabled the host organism to express proteins coded for by the inserted genes from a source organism. Genentech, the genetic engineering company co-founded by Boyer in partnership with venture capitalist Robert A. Swanson, developed one of the first commercially successful applications of rDNA. Human genes for producing insulin were inserted into *E. coli* bacteria, converting the bacteria into mini-factories for manufacturing insulin. By 1982, the pharmaceutical giant Eli Lilly began marketing the synthetic insulin, and the rDNA process largely replaced the older, more cumbersome method of extracting and purifying insulin from the pancreas of cattle or pigs. More abundant than the "natural" product and easier to produce, the synthetic version became the manufacturers' drug of choice for treating diabetes. Bacterial insulin effectively shut down animal insulin production in the United States, although some patients complained bitterly of worsened side effects. Yet, over the next twenty-five years the price drop predicted by biotech optimists failed to materialize, partly because the patented, recombinant version drove cheaper competitors out of business (Greene and Riggs 2015).

Expectations concerning who funds science, and thus who defines the aims of scientific research, were transformed in this period. In the aftermath of World War II, US scientists acquired a taste for generous federal research dollars, although many feared and resisted state control of research agendas. President Franklin D. Roosevelt, near the end of his life, asked Vannevar Bush, his chief science adviser, to come up with a workable model for future public funding of science. Evocatively

A New Lens on Life

titled *Science: The Endless Frontier* (1945), Bush's report delivered a blueprint for the future National Science Foundation (NSF) and a new social contract between science and society. American wartime research had gained mightily from brilliant, highly trained scientists fleeing fascism in Europe. Bush believed something had to be done to eliminate future dependency of that kind, but without turning science into just another arm of the state. His solution was to insist on a bright line between research and application. Public money would support disinterested "basic" research, leaving scientists relatively free to define their research goals without bowing to immediate governmental demands. In return, scientists would ensure the quality and significance of their work through unbiased peer review, supply the nation with trained personnel in times of need, and eventually produce applications benefiting society. Under this formulation, molecular biologists were at liberty to wander where their discoveries led them, unconstrained by governmental or taxpayer directives.

Bush's 1945 report had little to say on the means by which results would make their way into useful applications. Intent on justifying continued state support of science, Bush paid little attention to the role of private money or incentives. But by the time biotechnology matured into a distinctive research area in the 1980s (Bud 1993), Bush's social contract, centered in the imagined ivory tower of university labs, was becoming a thing of the past. Biological discovery looked profitable as never before. The elegance of the recombinant genetic technology aroused enormous excitement as financiers and manufacturers began betting on applications in medicine and agriculture. Academic scientists recognized that money was no longer just a resource for feeding young researchers or playing with fascinating problems for their own sake. There was money to be made, lots of it, from breakthrough technological uses. The unlocking of private funds for research and development meant that scientists began to construe their freedom not simply as a matter of choosing research projects, but also as an unfettered right to form profitable relations with the private sector. In an apt coincidence, Paul Berg won the Nobel Prize on the very same day in October 1980 that Genentech went public. Each partner's share in Genentech was valued at about $65 million (Cole 1980), no mean nest egg. Neither Herbert Boyer nor Stanley Cohen, however, pioneering toolmakers though they were, received the nod from the Nobel committee for modern science's ultimate accolade.

Particularly in the United States, the 1980s were the decade of biotech start-ups, as venture capitalists saw increasing promise in small, spin-off companies set up by academics hoping to get rich in a suddenly hot market for biological products. Federal policy propelled the gold rush. The 1980 Bayh–Dole Act required universities and nonprofits to patent and actively commercialize inventions from their federally funded research, but it also allowed them to retain ownership rights in these inventions.[4] A major row that shook the molecular biology community in 1992 shows how drastically their world had changed since the heady 1980s. In April of that year, James Watson abruptly resigned his post as the first head of the Human Genome Project (HGP) following a confrontation with Bernadine Healy, then director of the NIH. The NIH leadership had raised awkward questions about Watson's holdings in biotech companies and possible conflicts of interest, although his departure was more likely triggered by disagreements with Healy over the NIH's patent policy. Never one to mince words, an exasperated Watson bluntly told a *Science* reporter: "I don't know how to get someone to succeed me. I don't know anyone who doesn't have stocks. And I don't know anyone who would want to live with my boss" (Roberts 1992, 302). Clearly, by this time it was the stock-holding biologist, at ease with business dealings, who was in the driver's seat, and it was the scientist detached from the lure of the market who had become the endangered species.

Genomic futures

Biology, needless to say, was not the only axis on which science and technology progressed. By century's end, developments in computing and information science and technology linked up with the life sciences to produce further reorientations in biology's answers to the age-old question of where humanity is headed. The launch of the international Human Genome Project in 1989, with James Watson as its first US director, was one harbinger of change. Often described as biology's first "big science" project, the HGP was by many indicators indeed a massive undertaking. By the time the US National Human Genome Research Institute (NHGRI), one of the NIH's twenty-seven institutes, announced the complete sequencing of the human genome in 2003, the project had already attracted hundreds of scientists, collaborations among labs in six countries, and about $2.7 billion in public funds, although this was 10 percent less than the originally estimated $3 billion. With a reputation for fiscal responsibility, the HGP became a poster child for wise investments of public money.

The completed map of the human genome shifted biology's attention from a primary focus on genes to seeing entire organisms in informational terms, constituted as much by data as by the physical structures that carry the relevant codes. The sites of biological invention also shifted almost as radically as in the move from the field to the laboratory a hundred years earlier. The genomic transformation, as the science studies scholar Stephen Hilgartner perceptively writes, displaced the lab from the epicenter of knowing and doing: "Changes in policy paradigms, legal framings, and business models are also deeply implicated in creating new knowledge objects, reconfiguring old ones, and reordering the strategic objectives that these entities are used to pursue" (2017, 124). Here, the word "objectives" is key. It signals not only a displacement in the sites and modes of biological research, but also a change in what biology hopes to accomplish when addressing life's value and meaning.

In the shift from labs to a more networked production system, life became explicitly programmable and speed entered the discourse of biology. One marker was the rancorous race that broke out between the NHGRI and the maverick biochemist and bioengineer J. Craig Venter to see whose approach to sequencing would deliver results the fastest: the public project led by the NIH or by Celera, the company formed by Venter to get past what he saw as frustrating bottlenecks at the NIH. Competition eventually was declared to have been a good thing, delivering results ahead of schedule and under budget. In a post facto show of amity, NHGRI director Francis S. Collins and Venter appeared flanking President Bill Clinton on June 26, 2000 at a White House news conference announcing the publication of the first draft of the genome. UK Prime Minister Tony Blair attended via a video link, performing in a show of friendly internationalism.

President Clinton hailed the map's completion in terms that reprised Horace Judson's wonder at being in on the eighth day of creation. Clinton reminded his audience of Thomas Jefferson's dream of a continental map, one that Jefferson had seen laid out nearly two hundred years earlier in that very room and that "defined the contours and forever expanded the frontiers of our continent and our imagination." If maps are instruments of nation-building, as Benedict Anderson (1983) proposed in his pathbreaking work on nationalism, then this biological map in Clinton's vision infused a modern scientific project with decidedly theological overtones: "Today, we are learning the language in which God created life. We are gaining ever more awe for the complexity, the beauty, the wonder of God's most divine and sacred gift."[5] The implication was clear. There could be no fundamental division between scientific discovery and divine dispensation; life remained God's gift, and science helped by public largesse was learning to write in God's own vocabulary. Private money remained discreetly unmentioned, though Venter's visible presence hinted at its role behind the scenes.

To be sure, increased knowledge of our genetic makeup did not strike all hearers as an unmixed blessing, let alone a sacred gift. To some it brought fears of science out of control, as in Mary Shelley's 1818 fable of Dr. Frankenstein, or of states turning human life into an assembly-line commodity as in Aldous Huxley's *Brave New World* (1932). Others feared that genetic knowledge in the hands of employers or the insurance industry could lead to systematic exclusion of those with susceptibility to debilitating illnesses. Many more worried that, in societies where racial, ethnic, and gender classifications already correspond to deep inequalities, genetic knowledge would reinscribe the differences and authorize stigmatization and discrimination on newly strengthened scientific grounds. Confirming the worst of these fears, Watson himself caused a public furor and lost his position as Chancellor of Cold Spring Harbor Laboratory in October 2007 after granting a British newspaper an interview in which he stated that blacks of African heritage are less intelligent than whites of European origin (Hunt-Grubbe 2007).

Watson's fall from grace came soon after another cautionary tale for biological overreaching. In 2005, Harvard's president Lawrence Summers created controversy, and later also lost his job, for suggesting that "innate" differences between the sexes might account for the relatively low representation of women in the sciences (Traub 2005). Apart from the inappropriateness of such remarks from a prominent university president, Summers's call for more studies of this issue showed no recognition of the pathways by which biases often enter into research on links between biology and behavior. Extremely dubious conclusions about racial and gender differences were drawn in the past from studies using inadequately large or varied samples, insufficient controls on confounding factors, and very poor research management. Such studies were believed nonetheless because they confirmed dominant opinion. Research on human behavior is especially problematic because of the fuzzy nature of the thing being investigated, for example, in attempts to relate genes to "vagrancy," "curiosity," "violence," or "aggression." Genetic studies of behavior tend to reduce and essentialize complex phenomena; even maleness and femaleness are less well-defined and binary categories than research programs frequently take them to be. Critical readers of scientific hypotheses, especially on terrain as fraught as nature/nurture controversies, therefore need to question particularly closely the assumptions and conditions under which evidence of difference is produced.

Despite these well-founded concerns, well known to scholars who study science and technology, biology's claim to hold the master keys to human origins and differences remains very much alive. Harvard geneticist David Reich's book, *Who We Are and How We Got Here: Ancient DNA and the New Science of the Human Past* (2018a), offers an interesting contemporary example. Here, and in an op-ed in the *New York Times*, Reich (2018b) not only appropriated two of Gauguin's questions for genetics, but suggested that well-meaning commitments to equality should yield to new science, and that average genetic differences among populations need to be studied lest we lose valuable information about human variation. A distinguished group of sixty-seven scholars from the natural and social sciences, law, medicine, and the humanities acknowledged in an open letter on BuzzFeed that Reich had stated his claims carefully, but charged him nevertheless with ignoring the fraught social and political histories of words like race and population (see, for example, Reardon 2004), and even male and female.[6] Given the length, complexity, and persistence of these histories, especially in a time of resurgent xenophobia and racial tensions, what leaps out is Reich's conviction that all other ways of speaking about race should be held accountable to advances in science: "This is why it is important, even urgent, that we develop a candid and scientifically up-to-date way of discussing any such differences, instead of sticking our heads in the sand and being caught unprepared when they are found" (2018b). There is no suggestion here of a symmetrical awareness that scientific advances themselves might need to be contextualized against, and held accountable to, advances in other fields of thought and learning, such as ethics or law. In the next few chapters we look more closely at the frictions and complications that have resulted from this failure to acknowledge science's deep embeddedness in society.

2

Book of Revelations

In the twentieth century, science instrumentalized life in two ways, by learning to manipulate it and by profiting from its manipulation. First, as we have seen, lab instruments began probing the mechanics of life at ever smaller levels of organization, detecting structures and processes barely imagined by the nineteenth-century naturalists. These investigations also opened up possibilities for tinkering with DNA, the material stuff of life. Second, increasingly close alliances between science and private capital turned the knowledge acquired at the lab bench into instruments of gain, by engineering biological materials into new and useful entities for commercial purposes. Genentech's bacterial insulin was one of the earliest and biggest success stories, a poster child demonstrating that genetic knowledge could get past technological bottlenecks and make treatments more widely available, although the predicted price drop did not materialize as hoped. The practices of biology changed in the slide from field to lab and lab to market, as science became more tightly enmeshed with society's concerns: for example, through ties to medicine, information technology, agribusiness, finance, law, intellectual property, patient activism, popular culture, and the media. Craig Venter's race with the National Human Genome Research Institute to complete the sequencing of the human genome displayed the newly tangled relations between research for public good and for private gain.

These developments make it all the more paradoxical that the myth of science's institutional purity still retains such a powerful grip on the social imagination. Indeed, media commentators and even academics seem reluctant to connect the dots between scientific practice and its economic, social, and political supports except when scientists are charged with falsification or other misconduct. Money and politics, in short, are treated primarily as corrupting influences on a science that continues to be perceived and represented as a disinterested search for truth.

In this chapter, we explore the construction of this myth – of pure science's detachment from society's biases – by tracing two pathways through which scientific practice gets disembedded from its social matrix. First, scientists are their own most potent myth-makers. In influential firsthand accounts, discovery is remembered as a mental, nonmaterial process, fed only by imagination and genius. The reward structures of science reinforce these personal stories, often attributing insight to singular individuals, even though no less a figure than Isaac Newton wrote: "If I have seen further it is by standing on the shoulders of giants."[1] Second, science repeatedly asserts its right to be free from external controls because it is adequately self-policing and hence entitled to rule itself. Michael Polanyi, noted chemist and social theorist, forcefully argued for such autonomy in his 1962 essay "The Republic of Science." In his view science functions through a kind of continuous coordination among its practitioners, which he likened to Adam Smith's "invisible hand" of the market. In Polanyi's thinking, it is not that knowledge-making resembles the pursuit of commercial gain, but rather that the "self-coordination of independent scientists embodies a higher principle, a principle which is reduced to the mechanism of the market when applied to the production and distribution of material goods" (1962, 71). This assertion turns on its head the common picture of science as a marketplace of ideas. Rather, it is science, through its dynamics of free exchange and horizontal checks and controls among members, that serves in Polanyi's telling as the original model for the free market.

Polanyi wrote a decade or more before the advent of modern biotechnology, but views like his of a science that thrives on self-direction retain such a powerful grip on popular understanding that the science policy analyst Daniel Sarewitz called it "a bald-faced but beautiful lie" (2016, 6). If it is a lie, it is a lie with profoundly political consequences. Science, according to Polanyi, has a positive obligation to enlist public respect for its authority and independence: "Moreover, only a strong and united scientific opinion imposing the intrinsic value of scientific progress on society at large can elicit the support of scientific inquiry by the general public" (1962, 61). Biologists as a rule may not have read Polanyi, but their actions speak as if they had. At every major point of advance in the modern life sciences, the community has mobilized to affirm "the intrinsic value of scientific progress" and to resist government regulation as premature, ill-considered, and detrimental to society's interests.

Both strategies of detachment – from collective, cross-disciplinary enterprise to individual achievement, and from economic or political interest to the purity of curiosity-driven research – can be traced, in part, through the discourses of discovery in the sciences. We begin with some landmark moments in the postwar history of the life sciences, at each of which credit for unusual perspicacity was either claimed by or accorded to individuals rather than to institutions or collectives. Helped along by individualized rewards such as a Nobel Prize, this personification of scientific achievement contributes to the sense that science is driven by powerful minds that can decipher life's meaning on their own. Turning to science's demands for freedom from state supervision, we see how the presumption that discovery means progress, coupled to claims of effective self-policing, has warded off substantial controls on biotechnology, in spite of intensifying entanglements between research, politics, and money.

Biology's eureka moments

In a time when science has displaced religion as the authoritative source of natural law for many people, it is tempting to write the history of science as its own book of revelations: moments that separate our understanding of the world into a before and an after, the former a state of benighted ignorance, the latter one of light and knowledge. The Greek mathematician Archimedes gave us a name for such moments of enlightenment when, too excited to get dressed, he allegedly ran into the streets naked from his bath shouting "Eureka," or "I have found it." The "it" in question was how to measure the volume of an odd-shaped object like a gold crown. Archimedes had discovered by observing himself in his bath that solid bodies immersed in water displace equivalent volumes of liquid. The volume of the crown could therefore be accurately measured by submerging it in water and seeing how much water it pushed out of its way. The story may be apocryphal, but the word stuck, and scientists since Archimedes' time have racked up dozens of well-known, often self-proclaimed, eureka moments.

Fast-forwarding to modern times, Galileo's observation of the moons of Jupiter is often cast as a similarly revolutionary event in the shift from the Ptolemaic, or Earth-centered, to the Copernican, or heliocentric, view of the solar system, in which our planet is just one of several revolving around the Sun. Isaac Newton experienced his most famous revelation when an apple supposedly fell on his head, giving material heft to the as yet unarticulated theory of gravity. The German chemist August Kekulé said the ring-like structure of benzene, foundational to organic chemistry, fell into place for him during a daydream of a snake devouring its own tail. Marie Curie had her moment of brilliant inspiration when she noticed that the elements uranium and thorium give off the same amounts of radiation regardless of their compound forms; she deduced from this her transformative observation that radioactivity must be a property of atomic structures.

Fictive or well founded, accurate or grossly exaggerated,

the plot line of thrilling illumination, as instantaneous as the flick of a light switch, has proved irresistible to the makers and storytellers of modern molecular biology. Sometimes, the eureka moment gets coupled to the idea of a race, in which several scientists are competing to solve the same problem and the first to come up with the winning inspiration beats the others to the post. Other times, scientists' own misremembering perpetuates the notion of sudden insight, or it is amplified by acolytes and, today, the mass media – as it was in Archimedes' own day. Three thumbnail sketches from different phases of biomedical discovery in the twentieth century illustrate the moves by which discovery continually gets removed from its messy contexts to perpetuate the appearance of a science that feeds only on its own intellectual break-throughs and needs no other resources to flourish. These are the stories of penicillin, DNA, and genetic engineer-ing, each of which also says something important about the kinds of social supports science draws upon to capi-talize on its moments of insight.

Penicillin

One of the proudest tales in the annals of British bio-medicine is that of the bacteriologist Alexander Fleming, son of a Scottish farmer, who stumbled upon the anti-biotic properties of penicillin, it is said, on September 28, 1928, upon returning from a family holiday to his lab at St. Mary's Hospital Medical School in London.[2] Looking through his petri dishes, Fleming found one spot in which a fungus containing some mystery sub-stance, "mould juice," as he first called it, had seemingly kept the colonies of Staphylococcus bacteria from mul-tiplying, as they had in other dishes. "That's funny,"

Fleming laconically remarked as he showed the dish to a colleague (Lax 2004, 17). In that moment on a specified day, we are led to believe, the golden age of antibiotic discovery began, although the historical record offers a far more nuanced and long-drawn story, crossing decades, institutions, and continents.

Fleming's lab notebooks, to begin with, are sketchy, ambiguous witnesses. There is no mention of penicillin until October 20, and even then the timing of the scientist's five-week holiday does not match up with the two weeks he reported for the growth of the mold (Lax 2004, 18). For a while, Fleming himself did not know whether the mystery "juice" he found had weakened the bacteria or inhibited their growth.[3] In any event, although he identified the fungal strain and named the fluid "penicillin," he neither isolated it nor understood its disease-curing properties until much later. In 1940, working independently of Fleming, the pathologist and pharmacologist Howard Florey teamed up with the German-trained biochemist Ernst Chain to test penicillin's antibacterial properties on mice infected with the bacteria that cause puerperal fever in new mothers. Their results, published more than a decade after Fleming's first article on penicillin, caused great excitement in the scientific community, but industrial-scale production only became possible when a war and a transatlantic alliance created the incentives and built the infrastructure for the drug's widespread use.

Florey and Chain could not locate partners to develop their miracle find in a Britain already overwhelmed by the war effort, and so in 1941 they turned to the US Department of Agriculture's Northern Regional Research Laboratory (NRRL) in Peoria, Illinois to help scale up production. NRRL's longstanding work in developing fermentation media – and especially its use of corn-milling derivatives – dramatically increased the yield rate the Oxford scientists had achieved back home. Later still, US government advisers persuaded several major pharmaceutical companies – Lederle, Merck, Pfizer, Squibb, and eventually Abbott Laboratories – to take on the development challenge as part of their patriotic duty to serve the nation's military and defense needs. Together, these companies achieved remarkable results. Between 1943 and 1945, penicillin production rose from 21 billion units to more than 6.8 trillion units, and the price dropped from $20 per 100,000 units to less than ten cents (ACS and RSC 1999).

In 1945, Fleming, Florey, and Chain jointly received the Nobel Prize in Physiology or Medicine for their work on penicillin. All three were aware of each other's contributions and the enormous job of mobilization that had taken an unnamed fungal juice from a petri dish in a hospital lab to a savior drug for millions of lives. Enlisting American governmental science and corporate money was crucial to the effort, as was the collaboration between science and industry. Yet, in Fleming's own recollection, all that long, arduous effort apparently collapsed into one instant of glorious, accidental, game-changing discovery, with himself as the sole protagonist. He is widely quoted as saying: "When I woke up just after dawn on September 28, 1928, I certainly didn't plan to revolutionise all medicine by discovering the world's first antibiotic, or bacteria killer ... But I suppose that was exactly what I did."[4]

The birth of DNA

In its canonical version, the genetic age began just as abruptly in 1953, the year that James Watson and Francis Crick discovered the double-helix structure of DNA, the storehouse of information pertaining to the development, differentiation, and heritability of life. That biological organisms carry some mechanism, or "factors," for transmitting characteristics from one generation to the next had been known, as we saw in Chapter 1, at least since 1866, when Gregor Mendel published the results of his crossbreeding experiments. Two world wars and many social upheavals later, an international network of physicists, chemists, and biologists raced to unravel the structure of the material that carries the code of life. This was the puzzle that the 36-year-old Crick and his brash young American collaborator, the 24-year-old Watson, tackled in the University of Cambridge's Cavendish Laboratory.

The Double Helix (Watson 1968), Watson's breathless, irreverent, and idiosyncratically personal account of the discovery,[5] flouts convention in innumerable ways, perhaps most notoriously by reducing the story to one involving a very small cast of characters painted with almost cartoonish strokes. Watson emphasized the feel of being in a race with one of the giants in the field, Linus Pauling, a future two-time Nobel Prize-winner, and seen by Watson and Crick as the man to beat on the way to unraveling the structure of DNA. "Chiefly," Watson recalls in an opening scene set in Switzerland in 1955, "it was a matter of five people: Maurice Wilkins, Rosalind Franklin, Linus Pauling, Francis Crick, and me" (1968, 14). Of these, Wilkins and Crick shared the 1962 Nobel Prize with Watson;

Franklin had died of cancer in 1958 (a fact not mentioned in the book), and Pauling, as Watson gleefully observed, was defeated because, on the final lap, the eminent chemist made elementary errors in his chemistry: "If a student had made a similar mistake, he would be thought unfit to benefit from Cal Tech's chemistry faculty" (1968, 103).

Much of Watson's narrative focused on competing scientific teams racing to solve the same problem with different techniques and rules of interpretation. Franklin, the prickly, scientifically precise, and single-mindedly professional crystallographer, appears throughout the book as a character who defied the young Watson's understanding as a scientist as much as she eluded his sympathy as a woman and fellow human being. Repeatedly, Watson refers to his own lack of knowledge of crystallography and inability to make sense of writings and talks by the users of that technique, as opposed to physical model-building based on principles of structural chemistry. Nevertheless, a moment of great contention in the history of DNA is told as a moment of personal revelation. The trigger was one of Franklin's most information-rich images, the famous Photograph 51, shown to Watson by Maurice Wilkins when Franklin herself was not there to claim or interpret it. Looking at the photo, Watson, the crystallographic novice, had an epiphany: "The instant I saw the picture my mouth fell open and my pulse began to race. The pattern was unbelievably simpler than those obtained previously ('A' form). Moreover, the black cross of reflections which dominated the picture could arise only from a helical structure" (1968, 107). On this account, the image confirmed and ratified the structure that was already germinating in the minds of the Cambridge pair; it was simply the last bit of missing evidence.

Since the publication of Watson and Crick's article in *Nature* in 1953, historians have wondered how close Franklin was to reading her own image in the right way, whether she would have "got there" herself and how soon, and what kept her from a discovery in which her work with Wilkins is now seen as a crucial link. Crick's recollection of what Watson learned from Photograph 51 was distinctly underwhelming. Watson, he said, in a *Nature* article on the discovery's twentieth anniversary, had been shown the picture, "but he certainly had not remembered enough details to construct the arguments about Bessel functions and distances which the experimentalist gave" (1974, 766). It was not until Franklin and her colleague Raymond Gosling published their own article in the same 1953 *Nature* issue that, according to Crick, he and Watson became certain of the soundness of their idea.

In his introduction to *The Double Helix*, Sir Lawrence Bragg, the distinguished physicist whose group housed the Cambridge scientists, acknowledged – but glossed over – the ethics of according credit and authorship. His assessment is worth revisiting in full for that very reason:

> Then again, the story is a poignant example of a dilemma which may confront an investigator. He knows that a colleague has been working for years on a problem and has accumulated a mass of hard-won evidence, which has not yet been published because it is anticipated that success is just around the corner. He has seen this evidence and has good reason to believe that a method of attack which he can envisage, perhaps merely a new

point of view, will lead straight to the solution. An offer of collaboration at such a stage might well be regarded as a trespass. Should he go ahead on his own? It is not easy to be sure whether the crucial new idea is really one's own or has been unconsciously assimilated in talks with others. The realization of this difficulty has led to the establishment of a somewhat vague code amongst scientists which recognizes a claim in a line of research staked out by a colleague up to a certain point. When competition comes from more than one quarter, there is no need to hold back. This dilemma comes out clearly in the DNA story. It is a source of deep satisfaction to all intimately concerned that, in the award of the Nobel Prize in 1962, due recognition was given to the long, patient investigation by Wilkins at King's College (London) as well as to the brilliant and rapid final solution by Crick and Watson at Cambridge.

Quite apart from giving no word of recognition to Franklin's "long, patient investigation," the passage does its own discursive work to position the "DNA story," rather than the personalities who wrote it, in the center of attention. Importantly too for our purposes, the passage nicely illustrates one of the ways in which science arrogates to itself the functions of both judge and jury, through opaque internal processes that rank and reward merit. The Nobel committee, Bragg concludes, did right by science, apportioning "due recognition" to both the patient Wilkins and the more "brilliant and rapid" Crick and Watson.[6]

Biotechnology's breakthroughs
If science is thought to progress through bursts of individual insight, then one might expect technological

innovation, which almost by definition has to harness elements of the material world, to be a messier process, more aware of its dependence on things outside the minds of scientists. Yet, accounts of innovation in biotechnology equally stress the eureka factor. Herbert Boyer, who together with Stanley Cohen made seminal contributions to the technique of recombining DNA, is a prominent source of such stories. In a 2009 interview, the geneticist Jane Gitschier asked Boyer to look back on his "wonderful life" and to recapitulate its defining moments. Gitschier's classic fairytale opener "once upon a time" sets the tone, along with her framing observation that "one can almost define the revolution in molecular genetics by Herb's story alone." This will be a tale of individual genius. The interview as a whole tells a more complex and down-to-earth story. Chance encounters at conferences lead to momentous collaborations; technicians do the hard work of carrying skills and samples between labs and even across countries; and results build in slow increments from multiple, decentralized research initiatives. Yet, Boyer describes as personal revelations the moments at which he and his coworkers at the University of California discovered how the restriction enzyme EcoRI works. One such eureka moment finds him and a collaborator viewing the sequence of the cut sticky ends produced by EcoRI almost as if they are seeing a newborn child: "she and I looked at the results the next morning, a Saturday – it's GAATTC!" Another epiphany comes with the discovery that his lab can isolate any piece of DNA cut with EcoRI, regardless of where that DNA came from. In Boyer's own words: "That was another eureka moment. Bob Helling ... and I went to look at the gels in the darkroom, and there it was. It actually brought tears to my eyes, it was so exciting."

Other scientists, too, have spoken of their biotechnological breakthroughs in similarly reverential terms. Kary Mullis, the chemist who won a Nobel Prize for his pathbreaking work on polymerase chain reaction (PCR), the process used to produce thousands, even millions, of copies of segments of DNA, says the idea came to him one night driving down Highway 128 in California's wine country. That starlit inspiration on the open road has been described as a eureka moment "virtually dividing biology into the two epochs of before P.C.R. and after P.C.R" (Wade 1998). That is a narrative Mullis (1998) himself assiduously cultivated in his iconoclastic public performances. The sociologist Steven Shapin, reviewing Mullis's memoir in the *London Review of Books*, dryly noted: "At mile-marker 46.58 on Highway 128 – he had both the presence of mind and the sense of history to note the exact spot, if not the month – the epiphany arrives." The story, Shapin continued, "mixes the charm of heroic and crusty individualism with hoary cultural convention: Archimedes alone in the bath, Newton alone in the garden, Kekulé alone in his dreams" (1999, 17).

Other narrators, however, have complicated the storyline of brilliant epiphany. The Berkeley anthropologist Paul Rabinow carried out extensive interviews of the people involved in turning PCR into a research tool worthy of a Nobel Prize. These reveal a complex, human story of collaboration and conflict, involving the hands of technicians as much as the heads of scientists, so fractious at times that it led Mullis eventually to leave Cetus Corporation where he had begun translating his idea into application. In hindsight, "one could hold the view that all of the crucial work done to develop the value of PCR as a research tool and diagnostic method was done by others at Cetus after Mullis left, and that after 1986 he contributed nothing to this value" (Rabinow 1996, 133). It is only award committees and science journalists, Rabinow concludes, who "like the idea of associating a unique idea with a unique person" (1996, 161), even when the product owes its genesis to invisible, indispensable teamwork.

In keeping with the simplifications that gave biotechnology its fabled eureka moments, scientific citation patterns also subtly shifted to tie developments in biotechnology more strongly to the storied discovery of the double helix. The historian Bruno Strasser (2003) notes that the 1953 Watson–Crick article initially displayed the usual declining citation pattern for articles in the natural sciences, with a peak in 1963, the year the Nobel Prize was awarded. Starting in 1990, however, the year that saw the launch of the human genome project (HGP), citations began climbing steadily as HGP promoters, Strasser says, sought to tie their work to the double helix, whose image at the same time took off in popular culture as "the Mona Lisa of modern science" (Kemp 2003). This heroic genealogy elevated the HGP, sometimes dismissed as mere applied science, to the heights of fundamental research, animated only by science's thirst for discovery and divorced from the messiness of patent disputes and otherworldly concerns.

Yet, well before the 1990s, it was abundantly clear that American molecular biologists had more in mind than the sheer excitement of the scientific chase. By the mid-1970s, Genentech was already functioning as a private company lucratively spun off from university research; and the 1980 Bayh–Dole Act added more incentives for universities to encourage their scientists to profit from work in the lab. Research cultures on both sides of the Atlantic tilted toward privatizing the monetary benefits of inquiry. In the venerable University of Cambridge, UK, which celebrated its 800th anniversary in 2009, Tony Blair's New Labour government actively encouraged research scientists to form "spin-out" companies. Leading British scientists gladly embraced an opportunity that liberated them not only to pursue projects which could not be as easily undertaken in academic labs, but also to supplement their tightly controlled and comparatively meager university salaries (Jasanoff 2005, 240–1). Continental Europe, with its entrenched traditions of publicly funded education and research, was somewhat slower to follow the Anglo-American lead, but by the second decade of the twenty-first century, shifting cultural attitudes and government policies created a significantly more hospitable environment for start-ups in scientifically advanced countries such as France and Germany.

No other country, however, came close to the United States in the sheer wealth of economic and cultural capital accumulated by the pioneers of genetic engineering. Biologists holding academic jobs were suddenly rich enough to indulge their grandest dreams, as collectors of ancient Greek art or of eighteenth-century violins made by the Italian masters Stradivari and Guarneri del Gesù. The *Harvard Crimson* reported in 1995 that a prominent molecular biologist, Mark Ptashne, "also an accomplished violinist, businessperson and art collector," had tried "to convince then-president Derek C.

Bok to have the University loan him money to pur-
chase an early Stradivari violin." When Bok refused,
Ptashne crowd-sourced the funds by selling shares in
the instrument he craved to his friends (Chong 1995).
As the century advanced, biologists' fame spilled out of
scientific journals into the popular press. Kary Mullis,
perhaps the most unconventional Nobel laureate of the
modern life sciences, was featured – surfboard in hand
– in an *Esquire* magazine article (Yoffe 1994) titled "Is
Kary Mullis God (or Just the Big Kahuna)?"

And why not? In his review of Mullis's memoir,
Shapin remarked on the striking changes in the iconog-
raphy of the scientist since the postwar years. No longer
the world-weary sage or the emaciated ascetic, like an
Albert Einstein or a J. Robert Oppenheimer, Mullis, the
self-promoting scientist-entrepreneur, represented a new
breed of seeker after truth, happy to display his muscles,
his recreational preferences, and his appetites. Shapin
(1999, 18) asked: "Why should the entrepreneurial
scientist not want whatever the CEO of his company
wants? And why should he not now covet his neigh-
bour's ox, ass, house, wife or stock-options? No longer
an other-worldly and humble ascetic, the new-style scientist is finally free to be human, all too human." Just a
few years later, Craig Venter, the renegade NIH scientist
who nearly beat the federal government in the race to
map the human genome, garnered even greater fame
and visibility. Venter was invited to the White House,
featured on the cover of *Time*, and selected as runner-
up for *Time*'s person of the year. But fame vied with
fortune for the founder of Celera Genomics. An iconic
photograph by portraitist Gregory Heisler shows Venter
standing with arms crossed before a split black–white

background, looking confidently out at the viewer. His right side is dressed in a scientist's white lab coat and his left in a businessman's dark jacket. Commercial acumen, the image suggests, is now every bit as essential to the picture of a successful scientist as the thirst for discovery.

One can respect and appreciate the sincerity of the eureka stories told by scientists as accomplished as Fleming, Watson, Boyer, or Mullis, and yet their individually recovered memories are not accurate history. Rather, the inclination to produce and propagate such myths, even by scientists well versed in the accidents and complexities of their own research, has itself become a defining feature of modern science and technology. It is part of a culturally sanctioned repertoire of myth-making and credit-taking that allows biology to claim special authority in society's efforts to make sense of life.

Self-regulation 2.0: the Asilomar meeting

Eureka moments allow individual scientists to erase the work of teams and networks in knowledge-making, letting ideas shine forth as if they appear from nowhere. This is one potent argument for scientists to be let alone, to get on with their work unhindered. But the right to self-government also gets proclaimed at the institutional level, for the life sciences as a whole. Here, the basis for asserting independence from the rest of society goes back in part to claims such as Polanyi's about scientists' ability to spot flaws in each other's arguments and hence to hold their communities accountable to high standards of truth-telling. If science is itself a republic, then it follows that it is entitled to traditional forms of sovereignty, such as the right to set its own research agendas, establish standards of good practice, and police its own boundaries.

Two decades before Polanyi wrote his ringing defense of a republican science, the American sociologist Robert K. Merton provided a somewhat similar defense by observing that scientists subscribe to common ethical standards to safeguard science's institutional integrity (see Merton 1973). He identified four norms that constitute the ethos of science: communalism, universalism, disinterestedness, and organized skepticism. Unpacked, these terms mean that scientific knowledge is held in common; it is also invariant across contexts of use and application (universal), produced without regard for personal interests (disinterested), and held in check through systematic peer criticism (organized skepticism). That characterization of science still holds considerable public appeal, not least as an aspirational model for scientists, despite repeated demonstrations of a process that is in practice far more messy, contested, tribal, and driven by the rewards of the market.

Supplementing these generalized stories about the autonomy of science is a historical performance of self-governance that became an increasingly valuable reference point for American life scientists and biotechnologists from 1975 onwards. In February of that year, concerned leaders in recombinant DNA research, led by Paul Berg of Stanford, met together at California's Asilomar conference center to discuss under what conditions their experiments could safely move forward. For the preceding two years, molecular biologists had maintained a voluntary moratorium on potentially hazardous rDNA experiments, and now they believed it was time for that freeze to end. A specter hanging over the meeting, first identified by Robert Pollack of Columbia University, was that bacteria engineered with genes from the tumor-inducing Simian Virus 40 (SV40) might colonize the human gut and "unloose a plague" (McElheny 2003, 220). Pollack in fact likened the threat of runaway biological agents to atomic radiation, telling a reporter: "We're in a pre-Hiroshima situation. It would be a real disaster if one of the agents now being handled in research should in fact be a real human cancer agent" (Wade 1973, 567). A biological Hiroshima was too awful to contemplate. The thought that biologists, like physicists, might heedlessly let loose a destructive force comparable to the atomic bomb pushed activists in the molecular biology community to try to ensure that their science would proceed only under dependable, publicly approved controls.

Attended by about 140 participants, among them the barons and baronesses of a highly promising research field, the Asilomar Conference on Recombinant DNA developed an outsized reputation as a "meeting that changed the world." Recalling the event almost thirty years later, chief mobilizer Paul Berg noted with evident satisfaction that the "people who sounded the alarm about this new line of experimentation were not politicians, religious groups or journalists, as one might expect: they were scientists" (2008, 290). He approvingly described the risk-based regulatory proposals that resulted from it as a happy solution that allowed research to go forward and a thriving industry to develop without loss of public trust. The main lesson of

Asilomar, Berg – a scientist from one of America's premier private universities – concluded, was for "scientists from publicly-funded institutions to find common cause with the wider public about the best way to regulate." Otherwise the policy initiative would shift to corporate science and it would "simply be too late" (2008, 291).

Human recall, as we have already seen, is a slippery thing, and self-interested recall is more slippery than most. In recorded fact, the public was barely represented at that landmark Asilomar meeting, and a dispute soon erupted as to whose job it was to hold scientists accountable for the project on which they were so enthusiastically embarking. Horace Judson, the science journalist, chronicled a good deal of backstage grumbling among scientists who felt that the informal moratorium declared before Asilomar had kept them from doing valuable work. Sydney Brenner, a future Nobel laureate, was especially disconcerted by the attitudes of the American biologists. He told Judson: "I kept hearing them use the word *business*. You know, as in 'You'll put me out of business with those restrictions.' Many times" (Judson 1975b, 74). With feelings running high, there was considerable pressure on the scientists to come up with an arrangement that would satisfy both the firebrands and the cooler heads among the participants. Watson, one of the original signatories to the moratorium, was all for declaring an end to it and simply moving on. Nicholas Wade, writing in *Science* about the meeting, characterized him along with Joshua Lederberg as one of the *enfants terribles*, "breaking the ice for the faction among the younger scientists who were eager to get the moratorium lifted on the easiest terms feasible" (1975, 932).

Ultimately, the Berg group's conviction of the need for some restrictions prevailed, but it was a consensus that barely touched on the deeper meanings of human tinkering with genetic material. Fixated on safety, the conferees agreed on two kinds of controls, termed physical and biological containment. Physical measures would prevent the accidental escape of potentially harmful organisms, whereas biological measures would weaken novel strains so they would not easily survive outside the laboratory environment. Accompanied by a scheme for classifying future rDNA experiments into three broad risk categories – low, moderate, and high – these recommendations won overwhelming approval from the assembled scientists (Berg et al. 1975).

Asilomar's surprising legacy

To one pair of observant eyes, the Asilomar meeting seemed in its time relatively inconsequential – as if it might "rate at least a footnote in the history of science," even though it was a rare "example of safety precautions being imposed on a technical development before, instead of after, the first occurrence of the hazard being guarded against" (Wade 1975, 931). What happened, then, to turn an unprecedented but small and specialist scientific meeting into Paul Berg's meeting that changed the world? To understand that transformation, it is important to read the Asilomar rDNA meeting at more than one level – not merely as an exercise in precautionary regulation, but also as a public performance of Polanyi's republic of science, a government of scientists, by scientists, and for scientists. In that second reading, as J. Benjamin Hurlbut has shown, Asilomar reverberated through decades of advances at the frontiers of science as "Asilomar-in-memory," a precedent to be cited and continually relied upon as an example of successful self-governance whenever science feels threatened by restrictive state regulation. Asilomar became a trope for what Hurlbut (2015, 127) calls "an imaginary of governable emergence," the promise that scientists on their own have the capacity to take the uncertainties arising at any frontier of innovation and to make them "coherent, orderly, and controllable" (2015, 147). Such a promise in effect releases science from any need for external supervision as all plausible dangers can be foreseen and forestalled by scientists' collective wisdom. They are the only ones seen as competent enough to disentangle realistic scenarios from unrealistic ones and to tailor regulation to fit the "real" risks.

Not all of the original Asilomar participants, it appears, agreed with that reading. Judson reported that Brenner, speaking from his transatlantic vantage point, repeatedly insisted that it was up to society, by which Brenner meant those entrusted with science policy and funding, to determine how much risk was tolerable in relation to the projected benefits. Brenner also called attention to the asymmetrical distributive effects of research: immediate gains would be reaped by scientists, while more distant risks would fall upon innocent bystanders and even the world at large if something dangerous escaped from the lab. Even Brenner's imagination, however, was limited by his experiences as a researcher concerned with manipulating organisms at the molecular scale. How biological inventions might affect structures of equality or inequality in the world was not an issue for him, let alone worries about tampering with deep-seated religious or ethical understandings of the worth of life. The key to understanding Asilomar's success as a model of research governance lies precisely in the forms of imaginative (and, as we will see in Chapter 5, also discursive) narrowing that the meeting accomplished, reducing risks to only physical harms, while making it seem that a thoughtful scientific community had coped with all the possible negative consequences of emergence.

Many flashpoints that have arisen around the production and use of genetically modified organisms (GMOs) since the 1980s were bracketed and set aside, or not even brought to the table, at Asilomar. Most controversially, the deliberate release of modified plants for agricultural purposes was deemed too uncertain to be safely undertaken. When an NIH-appointed expert committee issued guidelines following up on the Asilomar recommendations, deliberate release was one of five categories of forbidden experiments. There were other omissions:

> Questions about biosecurity and ethics were explicitly excluded from the agenda. Ecological questions, such as long-term effects on biodiversity or non-target species, received barely a nod. The differences between research at the lab scale and development at industrial scales did not enter the discussion, let alone questions about intellectual property or eventual impacts on farmers, consumers, crop diversity, and food security around the world. (Jasanoff et al. 2015, 28)

It is not especially surprising that these wider issues fell outside the mental horizons of researchers whose eureka moments (recall Boyer's exhilarated cry, "it's GAATTC!") had focused before the Asilomar meeting mainly on precision work with restriction enzymes, plasmids, genes, and viruses in the lab. Nor is it so remarkable that, in 1975, corporate control of the life sciences was still largely off-screen; Genentech, the first significant biotech start-up, did not go public until 1980. It is more remarkable that Asilomar still stands as a monument to scientific self-governance despite the long years of global controversy precipitated by the allegedly well-governed recombinant technologies.

In the US context, Asilomar did its bit to defuse calls for comprehensive federal regulation. With the risk of labs releasing plagues seemingly under good enough control, momentum for state action shifted away from Congress back into the executive branch. Biotechnology, US policymakers concluded, was merely a new indus-trial process for making products that were already adequately controlled under existing law (Jasanoff 1995a). What emerged by the mid-1980s, therefore, was a relatively permissive regulatory environment for biotechnology coordinated by agencies stretching their mandates to incorporate genetically modified products. Contradicting in a sense science's own sense of discov-ery and novelty – its eureka moments – this mundane and domesticated policy discourse shelved, at least for a time, any concern that biotechnology's impacts might fundamentally trouble our sense of what life is, let alone what it is for.

Conclusion

Science is not only a source of wondrous discoveries about nature leading to life-enhancing technological applications. It is also a highly organized social activity, dependent for its very pursuit on many forms of buy-in and support. By the end of the twentieth century, big science was big business, and individual scientists were expected to do their part to attract substantial nongovernmental funding for their research. Yet, a considerable part of science's popular appeal still derives from claims of asceticism and individual genius that endow the discoveries, when they come, with a hallowed light, as if nature is speaking through anointed oracles. Few scientists are modest enough to say, as Francis Crick did in his 1974 reminiscence, that it is the discovery that makes the scientist and not the other way around: "Rather than believe that Watson and Crick made the DNA structure, I would rather stress that the structure made Watson and Crick" (1974, 768). In the case of DNA, Crick went on to say, it was the molecule's "style" that accounted for its exceptional impact. It was simple and beautiful, and it soon attracted large numbers of people and vast sums of money to exploit the opportunities for manipulation that the structure made possible. But this, too, is a radical simplification. It makes light of the convergence of scientists' self-interest, the ideology of deregulation, and the exclusionary dynamics of Asilomar that built an aura of inevitability around biotechnology's progress.

Here, as we have seen, a second form of myth-making enters the picture, the myth of science's social responsibility that has its own lineage in older commentaries such as those of Merton and Polanyi on the specialness of science. The Asilomar conference on recombinant DNA came to be seen as scientists moving proactively to regulate harms that might stem from their work, even promising to hold back from lines of research that, to their minds, present too much uncertainty or too great a probability of disaster. But that enactment of restraint came with costs for democracy. By seizing the initiative for deliberation at the dawn of a technological revolution, the molecular biology community established a virtual monopoly on framing the issues to be debated, the terms in which the debate would be conducted, and even to a large extent the legitimate objectives of governance. A focus on physical risks, originating with and limited to lab-created organisms, kept at bay wider social concerns emerging from law, religion, ethics, and culture. These concerns did not fade away from public consciousness, but the framework of containment laid down even before the Asilomar meeting continued to block reforms as societies all over the world struggled in subsequent decades to meet the ethical and political challenges of advances in genetics and genomics. That uneasy path dependency, with its multiple erasures, lock-ins, and evasions, is exposed to deeper critique in the following chapters.

3

Life and Law: Constitutional Turns

The transit from myth to history, the literary scholar Theodore Ziolkowski has written, was marked by the figures of great lawgivers, from Manu in India to Moses in Egypt (1997, 3). Law, in other words, appears as one of the first manifestations of the ordered, self-conscious collectives that we think of as societies. *Life* has enjoyed a complex and evolving relationship with law, and most especially with state power, ever since human beings began codifying their rules for living together. How to justify the right of a few lives to command many others, for example, is one of the most ancient concerns of the law. It dates from millennia before democratic publics asserted the right to govern themselves. Law defines the basis for kingly authority, the ruler's powers over the ruled, the responsibilities of government, and the duties of subjects toward the sovereign as well as each other. Most dramatically, the law decrees whose lives should be saved and whose forfeited when people have not lived within society's rules.

The Babylonian Code of Hammurabi offers one of the earliest examples. Hammurabi claimed the right to rule not only as a divine gift, strengthened by lineage and conquest, but also through his achievements as a maker and enforcer of law. Formulated around 1770 BCE, his code consisted of 282 provisions spelling out penalties for contract violations, assault and battery, sexual crimes, familial disputes, and common forms of bad behavior involving professionals from doctors to builders of ships and houses. Throughout the code, life hangs in the balance. Some rules stipulate punishments for taking life; others mandate death for specific violations of law. While endorsing the harsh ethic of an "eye for an eye,"[1] Hammurabi the code-giver also declares himself a protector and father of his people, a ruler who rights wrongs and makes sure the strong will not victimize the weak, the widow, or the orphan. Some have seen in these provisions the stirrings of a constitutional order, in which the sovereign takes responsibility for the security and prosperity of his subjects' lives.[2]

Other ancient legal traditions have also stressed the sovereign's duty to protect and promote life. Some five hundred years after Hammurabi, Moses brought down from Mount Sinai the stone tablets of law "written with the finger of God." The Ten Commandments or Decalogue and the long Covenant Code set forth in Exodus 20–23 served as the basis for the laws that have governed the Jews for more than three thousand years. The prohibition against murder – "Thou shalt not kill" – sits high on the list of commandments, marking a turn from piety (or Godward duties) to probity (or duty toward one's fellow humans). More generally, the laws given to Moses lay out principles of social order along lines similar to those in Hammurabi's law code. As told in the Bible, Moses himself chose to be a judge, showing his people the ways in which they should walk; later, following his father-in-law Jethro's advice, he became a trainer of judges. Moses in this way served not only as God's ambassador in proclaiming divine law, but also as the earthly administrator or ruler whose job it was to see that the law was faithfully executed.

A millennium later still, in the third century BCE, India's emperor Ashoka converted to Buddhism after winning a violent war. His attitude toward life changed dramatically. Edicts carved on rock faces and pillars throughout South Asia signaled the emperor's new-found faith by urging respect for all life. The Ashokan code mandated respect for parents, generosity toward priests and relatives, moderation in spending, and above all restraint in killing or harming living things (Thapar 1997). Comparable to the Hippocratic principle "do no harm," Ashoka's ethic of respect for life extended beyond humans to include all animals. Formerly, the emperor publicly confessed, hundreds of thousands of animals had been slaughtered for food in his kitchens, but hence-forth he would prohibit such wanton bloodshed. Even deer and peacocks would be spared. Prisoners sentenced to death were allowed slight reprieves: their relatives had three days to pursue an appeal. If no appeal mate-rialized, the condemned were given a chance to make amends through gifts and prayers, and so gain merit in the next world. These injunctions of benevolence and nonviolence, expressed in a strikingly personal voice, turned Ashoka into a "political fountainhead" for the leaders of independent India, who two thousand years later took up his imperial symbols, the wheel and the lion capital, as emblems for a newly imagined free and secular nation (Lahiri 2015, 13–14).

These examples reveal how religious beliefs and tacit theories of legitimacy were intricately interwoven with norms governing life in the earliest societies. Prohibitions against killing appear and reappear across centuries and continents, as does the idea that law must explicitly authorize any public taking of life. Sovereign power, even in these ancient, pre-philosophical texts, claims legitimacy through the preservation of subjects' lives, and by creating and maintaining the conditions for orderly, peaceful, social existence. In other words, it is not only the bare continuance of life that the great lawgivers cared for, but also its flourishing.

Modernity's settlements between life and law are more complicated, but no less essential for social order. Although care for life remains a central obligation of sovereign states, it is no longer the single, unifying, kingly figure – a Hammurabi, a Moses, or an Ashoka – who embodies the law and its proper administration. Modernity is a time of fragmentation and specialization, of delegated authority and esoteric expertise. In the modern state, life, both human and nonhuman, falls under the care of multiple entities with divided jurisdictions and divergent mandates. Expert agencies are entrusted with looking after selected aspects of a population's well-being: health, education, social welfare, environment, and security, to name but the most salient. Each domain is informed by its own forms of knowledge and styles of governance, with potentially inconsistent perspectives on the meaning and value of life. What counts as life and what is owed to it may be assessed and acted upon differently in health care and in environmental policy, for example. In the medical context, a body may be deemed so compromised that withholding treatment is the only viable option; environmental standards, by contrast, are often pegged to protect those who are at greatest risk of dying. What does remain constant in matters involving the governance of life today is something not known to the premoderns – namely, a mutual dependence between knowledge and norms, that is, between the discoveries of science and the dictates of law.

Given the salience of life to our very understanding of law, it is hardly surprising that the twentieth century's biological advances entailed immediate and far-reaching legal repercussions. And given, too, the enormous power that law and science both wield in society, it was perhaps a foregone conclusion that these encounters were not always harmonious. Indeed, humanity's efforts to adjust to the post-genetic era have led to repeated confrontations between law and science. The most significant of these, as traced in this chapter, have constitutional overtones: in the era of lab-created biological entities, who and what counts as a human subject? What entitlements do such subjects have vis-à-vis their genetic information? And which authorities, public or private, determine how to protect novel forms of life? Interactions between biology and law around these questions have helped define an unprecedented claim of sovereignty – that of science. A new discourse of independence seeks to preserve a space of moral and political autonomy for science. It presumes a clear line of demarcation between scientific fact-finding and legal rulemaking, and upholds that separation through expert practices. Most central to this discourse is the notion that science is entitled to set its own research agenda, and hence to chart the frontiers of human advancement, but almost as significant is idea of the "law lag." This phrase encodes a very particular sense of the hierarchical relationship between science and law as regards the protection of life: on this view, progress indisputably lodges with science, while law must constantly extricate itself from outdated principles to give science its due respect. Science produces what is new and worthwhile in our understanding of life; the law trails behind, updating our values to keep pace with altered knowledge (Hurlbut 2017, 140–6). The concept of the law lag thus carries with it, almost as a mathematical corollary, a linear notion of progress that places scientific discovery always ahead of norms-making, and thereby cordons off the domain of inquiry as if it can move ahead without interference from the domain of values.

In what follows, I first look at some key events that helped shape this set of understandings around modern biotechnology and at alternative imaginations of social order that were set aside in the process. These were not accidental developments. American biologists played an active role in shaping and colonizing the deliberative agenda on genetic engineering. Because of America's well-known penchant for litigation, key moves took place in the US courts during the 1970s and 1980s. The rules that emerged from these conflicts were national in scope, formally applicable only within the US research environment, but the resulting vision of how to govern biological innovation resonated well beyond the territorial limits of the United States. That vision endowed science with considerable autonomy to decide when and how research should be regulated, and a concurrent right to negotiate the terms of coexistence – in effect a new social contract – with other institutions of governance, most notably the law. This line drawing resolved to some degree the perceived problem of the law lag, but it did so in an unexpected way. Instead of waiting for the law to resolve society's value conflicts over the manipulation of life in slow, incremental steps, biology increasingly took it upon itself to preempt legal intrusions into the affairs of science. And legal institutions often went along with their own disempowerment.

The sovereignty of science

The 1975 Asilomar meeting is a convenient marker from which to trace the rise of the post-genetic biological order and its claims to sovereignty. The consensus reached at Asilomar did not overtly displace the law. Indeed, it had no immediate binding force. It was a gathering convened by and for research scientists, and although the statement they issued was both published and widely reported, it contained no mechanism for controlling what scientists actually did in their workplaces. Congress, however, proved reluctant to legislate, dissuaded by scientists' arguments that it was too early and potentially too risky to put the brakes on work of great public benefit. In the vacuum of legislative action it fell to the NIH, the funding body responsible for biomedical research, to pick up the challenge. With little prior experience in policing research, the NIH had to reinvent itself as an organ of deliberative democracy for managing the risks of biotechnology; the demarcations it drew, both conceptually and procedurally, created in effect a self-regulated space for the emerging biotechnologies. In this respect, Sydney Brenner's belief that funding policy is the means by which society should control science proved to be prescient, if not exactly consistent with theories of democratic government.

Donald Fredrickson, NIH director from 1975 to 1981, the most formative years for biotechnology research policy, recalled that the organization formed for oversight purposes, the NIH Recombinant DNA Advisory Committee (RAC), met for the first time on the very day that the Asilomar meeting ended. RAC had started life in 1974 as an informal consultative body – dubbed a "kitchen cabinet" – for the director, but following Asilomar the committee assumed de facto regulatory responsibilities by issuing guidelines for federally funded rDNA research (Jasanoff 1995b, 141–3). Its first task was to adopt "the provisional statement of the conference as interim rules for federally supported laboratories in the United States" (Fredrickson 1991, 283) – or, put differently, to give lawlike status to scientists' judgments on how they should be governed. In this way, the RAC took control of a space for political debate that most leading biologists were unwilling to cede to society at large. In his blow-by-blow history of the Asilomar meeting, Fredrickson relates how a group called Science for the People, "a grass-roots science watchdog organization" composed of bacterial geneticists and molecular biologists, presented an open letter calling for the moratorium to continue until provisions for wider public input were in place. Fredrickson related, without irony, "There was no formal discussion of the letter at the conference, however, and scientific presentations filled the morning" (1991, 280). Comprising signatories from Harvard, MIT, and Boston University, Science for the People was hardly a typical grassroots group, but for scientists chafing to get on with research after a two-year moratorium, further delays were unthinkable, even when the impetus to reconsider came from among their own kind.

In keeping with the NIH's role as a research funding body, the RAC was conceived more on the model of scientific peer review than of authentic public oversight. Initially consisting of only twelve members drawn from molecular biology and microbiology, virology and genetics, its membership broadened two years later to sixteen, to accommodate a wider diversity of scientific fields. The committee expanded still further to twenty in 1978, with lay members added for the first time (McLean 1997). LeRoy Walters of Georgetown University's Kennedy Institute of Ethics, and RAC member, wrote in his foreword to Fredrickson's memoirs (2001, xii) that this last increase occurred under "gentle prodding" from the Secretary of Health, Education, and Welfare, and that only "[w]ith this expansion [did] the RAC [come] to be seen as broadly representative of the spectrum of public opinion on recombinant DNA research."

Seen as representative by whom, though? Reading between the lines of historical accounts by scientists and other observers, one finds a fierce, if quiet, pitched battle for control that scientists generally won, and that law, in the sense of collective values expressed by a representative democracy, lost. Senator Edward M. Kennedy, Democrat from Massachusetts, was a forceful spear-carrier for legislation informed by broad public participation. He joined forces with Senator Jacob Javits, Republican of New York, in 1976 (a year after Asilomar) to convene a conference of some fifty participants at Airlie House in Virginia to rectify what these lawmakers saw as the biggest problem: "There are suggestions that the scientists may not appreciate adequately the public interest and its role in decision-making" (Culliton 1976, 452). The senators' concerns for public steering understated the case in the view of some participants. The distinguished philosopher Hans Jonas of the New School for Social Research put the point most starkly: "Scientific inquiry claims untrammeled freedom for itself" (cited in Culliton 1976, 452).

For all of its commitment to dialogue, however, the Airlie House meeting produced no action plan and no sense of urgency comparable to the debates at Asilomar, which were driven by scientists eager to get on with their research. This second attempt to frame a deliberative agenda engaged no interest group with enough drive and ambition to counterbalance the force of organized science. Consequently, it was the Airlie House gathering that remained an obscure footnote in the official history of a scientific revolution, whereas Asilomar attained near-mythic status.

Lawmaking efforts by Kennedy and his staff continued, with the limited scope of the NIH guidelines as a specific target. Kennedy introduced legislation that would have extended federal supervision to all research involving rDNA and not only the publicly funded fraction covered by NIH grants. Institutional science vigorously fought back. Fredrickson, an ardent supporter of self-regulation, wrote: "In an intensive reaction to this and other proposed laws, scientists and their organizations soon made strong appeals to the Congress. The ardor of the legislators for statutory regulation cooled progressively during 1977–78" (1979, 154). Law, to put it bluntly, did not simply *lag*; it was effectively put *hors de combat* by a science determined not to accept lawmakers' controls on the right to determine the directions of rDNA research. Researchers in this way arrogated to themselves the power to determine not only which questions were worth pursuing, but also how much scrutiny their work should receive from public authorities.

Trials in the fields of rDNA

The Asilomar scientists imagined risk mainly as the unplanned escape of pathogens from the laboratory, and the mechanism they chose to guard against such accidents was "containment." In July 1976, closely tracking the Asilomar conference recommendations, the RAC issued guidelines calling for three approaches to containment: standardized practices for handling infectious agents in microbiology labs; physical barriers, such as closed doors and exhaust and ventilation systems; and biological measures designed to reduce the capacity of gene-altered organisms to survive outside lab conditions. Five classes of experiments were prohibited outright as either not adequately containable or posing grave dangers if containment failed. Fourth on the prohibited list was the "deliberate release into the environment of any organism containing a recombinant DNA molecule."[3]

The ban on deliberate release proved especially irksome to scientists looking for profitable applications of rDNA techniques, since it effectively blocked promising agricultural applications of the emerging technology. Within two years, the RAC relaxed its guidelines to keep up with science's demands for relaxing the ban, now authorizing the NIH director to approve such experiments case by case (Jasanoff 1995b, 151). Two Berkeley researchers, Steven Lindow and Nickolas Panopoulos, came forward with a plan to help the valuable but perennially vulnerable California strawberry industry. A common strain of bacteria, *Pseudomonas syringae*, induces frost formation on crop plants, resulting in premature crop loss and reportedly costing American farmers some $1.7 billion each year in 1980s dollars (Associated Press 1987). Lindow and Panopoulos had engineered a gene-deleted variant to eliminate the bacteria's ice-nucleating properties. Marketed under the trade name Frostban, the engineered bacteria were expected to compete with and displace the more prevalent ice-plus variant, thus helping plants to survive snap freezes. The RAC approved the proposed experiment with little debate, but the Foundation on Economic Trends (FET), an environmental organization formed by the anti-biotechnology activist Jeremy Rifkin, sued to stop it. FET charged that the NIH had deviated from its own guidelines by failing to carry out the public environmental impact assessment required for an experiment of this kind.

The resulting legal judgments survive today as reminders of a rare moment in which emerging biotechnology failed to dissociate itself completely from legal supervision. In *Foundation on Economic Trends v. Heckler*, the United States Court of Appeals for the DC Circuit held that the NIH had indeed erred in approving the experiment without first considering environmental impacts. It had "not yet given adequate consideration to broad and important issues relating to its role in approving deliberate release experiments."[4] All three judges on the appellate panel concluded, though for somewhat different reasons, that the NIH had not properly explained why releasing larger than normal quantities of an engineered life form, albeit one found in nature, would not adversely affect the environment. Judge Skelly Wright called attention to the double-edged purpose of the environmental impact statement. It was not merely a device to let experts assess the environmental effects of powerful new technologies, but also a warrant for an informed public to review and comment on the results of that scrutiny.

This democratizing move of the law did not sit well with the framers of Asilomar's de facto constitutional convention on how science should be governed. In a revealing essay in a Yale law journal, Maxine Singer, a co-organizer of the Asilomar meeting, showed that she either did not or could not understand the principles of democratic reasoning that resonated so strongly with Judge Skelly Wright. To Singer, the issue was quite simply a matter of who knows best when research should be allowed to proceed. Since the issue was, for her, first and foremost a question of safety, it was completely appropriate to leave it to scientific judgment. Second-guessing by judges or publics simply made no sense.

Singer may not have been familiar with Michael Polanyi's argument that curiosity-driven research needs no external controls, but her views aligned well with his notion of the republic of science. Only, for Singer and most other molecular biologists, the all-important point was not that scientists cross-check each other's results, but rather that they know best what society should be worried about. In the ice-minus case, the appropriate experts had assessed the risks and found them too trivial for concern. It was an open-and-shut case. "As I have already pointed out, however," Singer wrote, "there was no reason whatever to believe that the proposed field test would affect the human environment at all, let alone significantly. Moreover, it is difficult to think that seeding a tiny potato patch with a relatively incompetent variant of a common bacteria is a major federal action" (Singer 1984, 333). At stake, however, was not the rightness of scientific beliefs about bacteria, potatoes, and the human environment. The legal controversy centered on the right deliberative relationship between science and society, on how expert beliefs that touch upon the governance of life should be debated in the public square. The "major" federal action challenged here was the act of moving an entire class of environmental experiments from the once-banned to the not-banned category. That move's political significance had nothing to do with the size of a "tiny" experimental potato patch. The issue was the big one of orderly governance and legally ordained democratic process, not of a trivial intervention with a scientifically valid risk assessment.

This distinction consistently eluded scientists steeped in the ethos of a sovereign science. Many US geneticists continue to believe, with Singer, that their early concerns, expressed in the rDNA research moratorium and the subsequent Asilomar meeting, were exaggerated and had needlessly fed public fears. As recently as 2013, Steven Lindow, co-inventor of Frostban, asserted in the abstract of a lecture on the history and regulation of agricultural biotechnology: "Public perceptions of such studies have led to lawsuits, community protests, and stringent, somewhat arbitrary regulation of field research."[5] This statement neatly captures the science-first argument. Here, judgment as to how much or what kind of regulation is appropriate rests, first, with the research community. If non-scientific publics disagree, then it is their views that are arbitrary and their perceptions that are the problem. How to win them over? Scientists sharing in this mindset have often demanded re-education, a kind of social engineering, for lawyers and judges to bring legal outcomes into closer alignment with science's conclusions. To correct what she viewed as an absurd judicial ruling on ice-minus, Singer herself proposed a new literacy test for the legal profession: that "lawyers be required to display a basic knowledge of biology and other sciences," and even that the LSAT, the standardized admissions test for American law schools, be reconfigured to include "questions concerning science" (1984, 334). Rarely, if ever, have calls run the other way, asking for more training in law and democratic theory for scientists, as if knowing science is by itself a sufficient basis for knowing how best to govern it.

A royal abdication

In the decades since the ice-minus controversy, agricultural biotechnology has engendered worldwide resistance and become almost a by-word for mismanaged technology policy (Jasanoff 2016, 87–105). It would be wrong to think, however, that US courts, following *FET v. Heckler*, embraced an expansive conception of the public's right to deliberate on alternative visions of nature, environment, or life itself in controversies involving genetics. The story, rather, is one of growing scientific confidence and judicial retreat. One finds throughout the record of law and policy in the 1980s an acceptance of the promises made by science and a desire not to put brakes on claims of technological progress – sanctioned by a constitutional order committed to the promotion of science and "useful arts," or what we would today call technology. In 1980, the year that swept Ronald Reagan and his small-state, deregulatory ideology to power, the US Supreme Court articulated this hands-off vision in a landmark case that affirmed the patentability of living organisms. Less obviously, that case also reinforced an extremely liberal interpretation of the sovereignty of science.

Much discussed and much dissected, *Diamond v. Chakrabarty* on its face raised a simple question of statutory interpretation: is a living organism patentable under the applicable federal law? But deeper matters were at stake. To what extent should technologically altered life be treated as capital, and who decides? In a split 5–4 decision, the Court ruled for the interests of science and industry, moving away from its public-leaning ways of the 1960s and largely accepting the arguments set forth in an *amicus* brief by Genentech, the start-up firm co-founded by Herbert Boyer. The majority interpreted the 200-year-old law as granting a very broad right to patent "anything under the sun that is made by man,"[6] a reading that clearly covered humanmade micro-organisms that did not exist as such in the wild.

For our purposes, however, what stands out as most significant is the Court's denial of power on the part of either the legislature or the judiciary to resist the advancing tide of science. Writing for the majority,

Chief Justice Warren Burger observed:

> The grant or denial of patents on micro-organisms is not likely to put an end to genetic research or to its attendant risks. The large amount of research that has already occurred when no researcher had sure knowledge that patent protection would be available suggests that legislative or judicial fiat as to patentability will not deter the scientific mind from probing into the unknown any more than Canute could command the tides.[7]

The image of a helpless King Canute works here as a metaphor for the law's utter lack of agency with respect to scientific developments. Against science's inexorable tide of advancing knowledge, law is cast as merely obstructionist and arbitrary. These few lines set forth a vision of the relationship between law and science that is constitutional in a sense that the Court may not have expressly intended. Characterizing both legislative and judicial steering of science as rule by "fiat" (that is, by unassailable decree), the majority essentially rejected the possibility of legitimate democratic controls on scientific progress. It disavowed the notion that the public will, expressed through constitutionally ordained processes, *should* take precedence over what Hans Jonas referred to at the Airlie House meeting as scientists' "untrammeled freedom."

Diamond v. Chakrabarty can be read, then, as a separation of powers decision on two levels. First, as a formal matter, the Court deferred to Congress, by accepting what it took to be the legislature's intent with regard to the scope of life patents. But, second, it also drew a bright line between the internal dynamics of science and the law's capacity to impose meaningful restrictions on them. Here, the Court in effect created (or underwrote) a contemporary constitutional order in which the workings of nature and of society are subject to different pressures, subscribe to different priorities, and call for different social controls. In the Supreme Court's imagination, as articulated in *Chakrabarty*, science, as the institution concerned with unveiling nature, operates according to its own sovereign urges, rhythms, and desires that are simply not amenable to management by other institutions. Even society's legitimate concerns, such as how to allocate the rewards of discovery, must yield to the first-order imperative to respect science's autonomy: the law's most pressing duty, from this perspective, is not to impede the progress of research.

By giving free rein to scientific advancement, the Supreme Court passed up an opportunity to throw the question of appropriate limits on life patenting back to Congress. Questions about which features of life can be capitalized, and hence owned, could have been subjected to a broader, more inclusive debate in the 1980s if the Court had ruled as the minority advocated. Instead, those issues retreated into the turbid backwaters of legislative and executive politics, particularly in disputes regarding human embryonic stem cell research – as discussed in the next chapter. Some juridical developments in the twenty-first century, however, illustrate how the law's foundational commitment to upholding the natural order of things continues to perturb the hardline boundary between science and law drawn in *Chakrabarty*. These examples suggest that modernity's constitution is not irrevocably tied to quite such a singular, and deferential, understanding of the sovereignty of science as Chief Justice Burger took for granted.

Counternarratives: natural law

In *Chakrabarty*, the law abdicated from its enthroned power to declare how life should be valued in deference to science's allegedly superior right to disclose, and by extension to manipulate, the machinery of how living nature works. But the law's self-understanding in relation to nature and natural life is a more complex affair than this decision alone suggests. Curiously, it is in the murky thickets of patent law that a more nuanced vision has taken shape of limits on science's power to determine what life is worth. Three contrasting decisions, two from common-law courts and one from the context of biomedical research, speak to a cartography of sense-making in which the powers of science and law to set limits on human appropriations of nature were divided on more equal terms than in *Chakrabarty*.

The first is a decision by the Canadian Supreme Court in 2002 to deny university researchers from Harvard the right to patent the so-called oncomouse, a standardized test animal genetically configured to make it more susceptible to contracting cancer in lab studies.[8] The desired patent had already been granted in the United States, where the gene-altered mouse easily met *Chakrabarty*'s standard for patentable subject matter: "anything under the sun that is made by man." The Canadian court was asked to determine the issue of animal patentability under a law worded almost identically with the US patent law (the only deviation being that Canadian law retains the word "art" along with process, machine, manufacture, and composition of matter). By a narrow 5–4 majority, the Canadian court drew a line at what it called the patentability of "higher life forms," holding that these are not explicitly included in the definition of "manufacture" or "composition of matter," and hence do not count as inventions for patent purposes.

At one level, this too was a simple separation of powers decision, in that it drew a line between legislative and judicial authority. The court addressed the locus and kind of deliberation that would be needed to authorize a perceived major turn in the law, such as widening the scope of patents to include an entirely new category of inventions. The Canadian Parliament, the majority observed, had not defined "invention" as "*anything* new and useful made by man" (emphasis added). Invention therefore was not a completely open-ended category for patent purposes, unlike what the US high court had held. The justices concluded that, given the lack of specific instruction in the law, it was up to Parliament and not the courts to decide whether a higher life form, even one that had been genetically modified and hence was not found in nature, could be construed as a patentable invention.

More interestingly, the majority offered a metaphysical disquisition on the difference between natural and artificial life to justify its misgivings about expanding patent law to include all lab-created life forms:

Although some in society may hold the view that higher life forms are mere "composition[s] of matter," the phrase does not fit well with common understandings of human and animal life. Higher life forms are generally regarded as possessing qualities and characteristics that transcend the particular genetic material of which they are composed. A person whose genetic make-up is modified by radiation does not cease to be him or herself. Likewise, the same mouse would exist absent the injection of the oncogene into the fertilized egg cell; it simply would not be predisposed to cancer. The fact that it has this predisposition to cancer that makes it valuable to humans does not mean that the mouse, along with other animal life forms, can be defined solely with reference to the genetic matter of which it is composed.[9]

Critics complained that the distinction between higher life forms and others is vague and difficult to sustain in biological terms. Others have simply noted that patents on genes may render the need for patents on organisms irrelevant. They cite particularly a 2004 decision of the Canadian Supreme Court on a patent infringement claim by Monsanto against a farmer named Percy Schmeiser. Here, the Court ruled that Schmeiser, by consciously cultivating and replanting seeds from plants resistant to Monsanto's Roundup herbicide, had violated the company's property rights in the altered genes that carry the resistant trait.[10] In effect, the Schmeiser decision allows the informational content of genes to be patented, even if the organism containing those genes cannot.

These technical arguments concerning the scope and purpose of patent protection sidestep the question of interpretive authority that most concerns us here. By denying a patent on the oncomouse, the Canadian high court imposed two kinds of restrictions on science's right to define the limits of power to appropriate life. First, the decision reaffirmed that in complex and ambiguous territory relating to the control of life, rules must be democratically authorized, in other words, by parliamentary action. Second, the Court created a space for "common understandings of human and animal life," which hold that, for "higher life forms," the essence of life cannot be reduced to "the particular genetic material" of which it is composed. Lay views about the meaning and worth of life, in other words, were given parity with the views of science in this historic case.

To the surprise of most patent law experts, the US Supreme Court itself retreated from *Chakrabarty*'s *carte blanche* for life patents in a 2013 decision challenging patents on human genes.[11] That case questioned the validity of patents held by Myriad Genetics on genes known as BRCA 1 and BRCA 2, used as diagnostics for increased susceptibility to breast and ovarian cancer. Myriad claimed a right to patent these genes because the company had isolated them and thereby enabled their use in a purified form not found in nature, that is, within the human body. Myriad's position drew support from decades of prior practice. The US patent office had agreed that isolation and purification of genes were sufficient to move these biological materials out of the domain of natural objects into that of patentable inventions. Accordingly, when the American Civil Liberties Union (ACLU) decided to question Myriad's patent rights, few thought the challengers stood any chance of winning.

Confounding the skeptics, the Supreme Court ruled in a near-unanimous decision that human genes are not patentable.[12] To get there, the Court applied the "product of nature" doctrine, arguing that Myriad had discovered nothing that was not already "naturally occurring": "Myriad did not create anything. To be sure, it found an important and useful gene, but separating that gene from its surrounding genetic material is not an act of invention." Although the Myriad decision marked a significant step back, many questioned whether it would have much impact on the biotech industry, since the Supreme Court expressly exempted complementary DNA (cDNA) from the scope of its ruling. Unlike genes themselves, these synthetic sequences, the Court noted, do not occur in nature because they consist of only the coding regions of the DNA that direct the formation of proteins. By holding patents on cDNA, companies may therefore still assert property rights over the all-important informational content of genes, rendering the control of entire genes superfluous.

From a constitutional perspective, however, the Myriad decision, like the Canadian oncomouse decision a decade earlier, reinserted the law into a debate on the meaning of life from which *Chakrabarty* had notably retreated. It curtailed the power of the life sciences and technologies to blur distinctions that human societies have made since the beginnings of time between what we take to be natural and what can be claimed as products of human ingenuity. In defining a class of living things – higher life forms – that fall outside the scope of patentable subject matter, the Canadian high court affirmed a boundary that scientific enterprise is not entitled to cross without explicit legislative permission. Similarly, in the Myriad controversy, the US Supreme Court affirmed that naturally occurring entities cannot be claimed for private gain simply because someone has done the work to make them more available for subsequent use – as in the case of isolated human genes.

The third example also carried constitutional implications, though without formal recourse to the law. This is the now-famous case of Henrietta Lacks, a young, black, American woman who died of ovarian cancer at the age of only thirty-one in 1951. At that time, Lacks was under medical care at the Johns Hopkins Hospital in Baltimore, Maryland. Her physicians succeeded in extracting cancerous cells from her dying body and, by getting them to reproduce indefinitely in the lab, created one of biology's most widely used research tools, the HeLa cell line. The letters of her name lived on, attached to a lab-created artifact and imparting to Lacks an undying if anonymous fame; yet, the relationship between the cell line and the woman it came from was essentially forgotten. Forgotten, that is, for sixty years, until an enterprising investigative journalist, Rebecca Skloot, resurrected the sadly truncated history of the living woman in her best-selling book, *The Immortal Life of Henrietta Lacks* (2010).

Skloot's story sent tremors down the fault lines of American biomedicine and racial politics. It brought under a single banner of moral outrage the historic subjugation of enslaved black bodies, the bioethical horror of experiments with black patients in the Tuskegee Syphilis Study, the persistent neglect of health disparities between blacks and whites in America, and the arrogance of a biomedicine that seems at times to care more for its own advancement than for the lives entrusted to its care. Actions by researchers in the European Molecular Biology Laboratory (EMBL) in Heidelberg, Germany added fuel to a smoldering fire. In March 2013, an EMBL team led by Lars Steinmetz announced it had sequenced the HeLa genome and was ready to publish it. That information pertained not only to Lacks herself but to descendants who shared her genetic traits, and yet no representatives had been consulted about EMBL's publication plans. Infuriated, and newly endowed with voice and visibility, the Lacks family demanded redress. As the entity responsible for funding American biomedical research, the NIH recognized that something had to be done. In an ad hoc but pragmatic solution, the NIH granted members of the Lacks family a permanent seat at the table in any future decisions involving research with the cells derived from their ancestor's diseased body.[13]

Belated and partial, and concluded without huge fanfare, the Lacks cell-line settlement nevertheless demonstrates the power of alternative constitutional readings of both what lives are and what they are for. It asserted, in the first place, that our cells are us, at least to the extent that we have a tacit right to say if extracts from our bodies can be used for research. But the settlement also sent a message about how far science could go in using human biological materials without regard for concerns other than its own.

Under the powerful eye of twenty-first-century racial awareness and demands for racial justice, demands that biology has done a lot to legitimate, the Lacks story could not be let stand as validating science's moral right to define the social good in keeping with its own imperatives. In this case, there was no doubt that much beneficial research had been done with the HeLa cell line or that such work should continue. At the same time, by giving the once-silenced Lacks family a role in approving future research, a national scientific agency affirmed a kind of constitutional responsibility to ensure that a patient's subjectivity does not get erased just because she is poor, black, ignorant, sick, or even dead. Henrietta Lacks died unsung, not knowing that her cells would outlast her mortal body. By granting her biological kin the right to represent her in research using those cells, the NIH gave Lacks – and her right to privacy and autonomy – a more embodied immortality than the living woman could ever have imagined for herself. Perhaps most strikingly, even international science was required in this instance to hold back in deference to human values articulated in the United States.

Conclusion

Engagements between biology and the law, especially in the United States, display some of the most significant moves by which the life sciences have arrogated to themselves the right to determine what life is for – a right that the law since ancient times regarded as its own grand prerogative. That preemption did not happen all at once, explicitly, or in ways that were apparent to watchful publics. Through repeated encounters, however, a powerful narrative took hold that cast science as society's most progressive force and anything that gets in science's way as backward and resistant to emancipation and enlightenment.

At several critical junctures when legislatures or courts seemed inclined to restrict the directions of research, scientists invoked what amounted to a transcendental public good argument – namely, that enhanced scientific knowledge of life is of supreme benefit to society and hence research should not be stopped or slowed under any circumstances. At Asilomar, in the RAC, and in controversies over succeeding decades, biologists claimed the right both to frame the issues of public concern and to offer solutions for addressing them. Legal institutions adopted, at times without question, biologists' and the biotech industry's narrowing of those issues largely to questions of physical safety and harm; and science's determinations of safety were deemed sufficient to stave off other arguments in favor of regulation.

In the resulting division of power between law and science, the "should" and "ought" concerns of the law were inevitably relegated to the sidelines, as concerns that must follow – or lag behind – developments in research. Yet, the "law lag" was never a matter of intrinsic institutional incapacity. As *Chakrabarty* forcefully demonstrated, this belittling of the law's role as a progressive instrument was more a mutually enacted construction, embedded in deep-seated cultural commitments to the advancement of science, shared and reinforced by the elites of both science and law. Moments such as these call into question the competence of institutions other than science to evaluate the costs and benefits of inquiry. The fact that the law so frequently accepts that frame, and restrains itself in struggles for control, then becomes further evidence of a laggard institution, lacking moral authority to adjudicate life's purposes or to steer the engines of progress.

Yet, determined citizen action can bore holes into this simplistic allocation of rights and responsibilities and force debate on issues other than those that science has framed as worth discussing. Thus, the RAC's inadequacy as a democratic forum was put on display when a US federal court insisted in *Heckler* that the decision to permit deliberate release of GMOs should be subject to a public assessment of environmental impacts. Scientists may blame such inconvenient forms of muscle-flexing on the scientific illiteracy of citizens and judges, but such cases should be seen instead as rare affirmations that societies can indeed reflect more deeply on the brakes they wish to impose on scientific advances – not to deny the merits of research, but to assert values beyond science's drive to reveal life's workings and remake the natural order. In this respect, the Canadian oncomouse decision, the Myriad gene patenting case, and the HeLa cell-line dispute all hold lessons that are worth recalling – as foundations for a more balanced and respectful separation of powers between science and law. Each of these cases, however, came about through specific, localized initiatives, and not through systematic inquiry into the purposes of biological manipulation, or answers to what life is for. To move from episodic, interest-driven challenges to continuous reflection and monitoring of biology's sovereignty claims would require democratic societies to acknowledge the slippage of power from law to science, and to strive for ways to reassert the normative prerogatives of the law.

4

Life in the Gray Zone

At the 1976 Airlie House conference to consider the relations between biomedical research and the public, the philosopher Hans Jonas remarked on the way the winds of change were blowing for the life sciences at the dawn of the genetic age:

> Science has its tasks increasingly defined by extraneous interests rather than its own internal logic or the free curiosity of the investigator. This is not to disparage those extraneous interests nor the fact that science has become their servant ... But it is to say that the acceptance of this functional role ... has destroyed the alibi of pure, disinterested theory and put science squarely in the realm of social action where every agent is accountable for his deeds. (Cited in Steinfels 1976, 22–3)

Not only were science and commercial interests developing greater stakes in each other's activities, with science thriving on "intellectual feedback precisely from its technological application," but this interdependence also meant, in Jonas's view, that scientists had become accountable for work that was increasingly instrumental, interventionist, and social. Pure science, the quest for truth and order for their own sakes, was a thing of the past, and so was the notion that nature's discipline by itself would ensure the responsible conduct of research. The mix of material manipulation, design, and external incentives that drove and rewarded the modern life sciences also allowed biologists to claim ownership of life in unforeseen ways. Life in bioengineers' hands assumed new forms that spilled out of existing conceptual categories. The entities emerging from the lab could not easily be classified as living or nonliving, human or nonhuman, expendable matter or beings entitled to moral respect. As ambiguous constructs, they defied society's established modes of sense-making. In this chapter, we trace some of the most important episodes of public debate around efforts to regulate engineered biological objects that fell into such moral and political gray zones. These cases display complex entanglements between judgments about how to characterize novel life forms (what life is) and prior social understandings about how life ought to be nurtured and governed (what life should be for).

The settlements that emerged from these struggles rested on different forms of deliberation and reasoning, and they were not uniform across countries. Comparison across decisionmaking cultures underscores the ambiguities and inconsistencies that accompanied biology's turn from analyzing life to remaking it. These contestations reveal in fine-grained detail the multiple meanings accorded to life – its nature and its purposes – in a world where life is seen more and more as matter to be perfected, or even transformed, through science's technical prowess.

Engineering ambitions

To understand contemporary biology's claim to be the primary custodian of life and its value, one has to set the turn toward bioengineering in a wider historical context. Modern biology, as we have seen, advanced through a series of moves that reflect changing understandings of what life is and how to investigate it: from the field to the lab, from the molecule to the genome, and most recently – perhaps most significantly – from the bench to the living body. The partnering of the life sciences with technologies for manipulating living entities gave rise to a new mode of biological sense-making, not merely by deciphering hidden genetic codes but by putting the material components of life together in novel ways. This engineering approach still took DNA as one of its primary raw materials, but biology's technoscientific ambition turned toward building new entities, instruments, and tools.

As with any radical changes in disciplinary agendas, the mash-up of biology with engineering can be traced to different starting points depending on the particular stories one wishes to tell. The choice of origins and key moments reflects the preferences of the history writers, who construct narratives in accordance with their personal inclinations as well as their own time's intellectual and social preoccupations. Thus, a historian inclined to trace patterns of ideas and practices back to their roots might look for the earliest signs of a particular way of working, as when biology first became industrial through use of fermentation technologies (Bud 1993). Others might sooner take their inspiration from changing phenomena in society. For example, when the feminist movement put the "woman question" on the table of scholarly inquiry, histories of women and gender in science were quick to follow (Rossiter 1984; Keller 1985). Similarly, by the late twentieth century, a wide-ranging interest in accounting for the material and humanmade dimensions of the world produced a new emphasis on what the biologist and historian Hans-Jörg Rheinberger called "epistemic things." He defined these as scientific objects, or entities "whose unknown characteristics are the target of an experimental inquiry" (1997, 238).

The material "things" of life, such as the transfer RNA and messenger RNA molecules at the center of Rheinberger's history, were not, however, mere playthings of scientists in labs. Living entities and their components carry meaning in all sorts of other social contexts, as plants do to farmers, pets to children, tumor cells and MRI scans to people with cancer, and the fetus in the womb to expectant parents. Changes in how we make or manipulate, or even describe, lab-created living objects necessarily perturb society. One story of human tinkering with life that proved profoundly consequential began with just such an entity of diverse social meanings, an artificially generated object that nevertheless encoded fundamental aspects of human life, and in so doing radically altered both what we think life is and what it is for. This was the product of efforts in a Manchester clinic to overcome problems of human infertility by making human embryos outside the female body.

Pioneered by two British medical researchers, the developmental biologist and eventual Nobel Prize-winner Robert Edwards and the gynecologist Patrick Steptoe, the technique of *in vitro* fertilization (IVF) reconstituted the boundaries of the natural in one of life's most basic processes – human reproduction. Through these two researchers' labors, a previously unseen and largely unexamined form of life, the early-stage human embryo, became a regular denizen of labs, an object to be made, tested, studied, treated, and eventually used as a source of various kinds of derived biological materials, most notably human embryonic stem cells (hESCs). Of course, IVF embryos also continued to serve their originally intended function, as proto-human lives to be implanted into the wombs of women unable to conceive without technological assistance.

The making instinct led to further advances in reproductive biology, most famously the first cloning of a mammal, a sheep named Dolly, from the cells of an adult mammal at Edinburgh's Roslin Institute in 1996. Dolly was the generational daughter as well as the biological twin of the mother from which her genome derived, a genome given bodily form and substance by transferring it to an egg cell, from another sheep, that was allowed to develop as if it had been fertilized by natural means.[1] Within two more decades, similar techniques would be used to create human embryos free from defects carried in the mother's mitochondrial DNA. To make these "three-parent embryos," a fertilized nucleus from the egg cells of a mother with a known mitochondrial disease would be substituted for the nucleus of a healthy, fertilized egg cell from another woman. The resulting embryo, containing DNA from two maternal donors, would then be carried to term like any other IVF baby.[2]

As discussed in Chapter 3, living bodies can be mined for cells of interest to science and turned into colonies, or cell lines, that can be indefinitely nurtured in lab conditions. The HeLa cell line was one of the earliest and, from a research perspective, most productive of such instruments. With the creation of embryos in labs, however, scientists found a new biological source from which to extract cell lines for research. In 1998 James Thomson of the University of Wisconsin at Madison became the first to isolate human embryonic stem cells from blastocysts, an early-stage but still immature embryo with differentiated inner and outer cell layers. These hESCs have a property that renders them more therapeutically useful than cell lines derived from adult human bodies. They are *stem* cells, that is, not yet specialized for turning into hearts, lungs, or kidneys, nerve fibers or skin surfaces, blood or brain. Such cells are called pluripotent; they can be programmed for a multitude of purposes, as transplants to replace and regenerate any kind of diseased tissues that might otherwise lead to the sufferer's irreversible debilitation and death. Like IVF embryos and cloned mammals, hESCs provoked intense questioning in societies uncertain how to position these objects morally in relation to actual or potential human lives. Derived from discarded embryos to heal living humans at risk of dying, hESCs seemed to pack within their small physical compass all three of Gauguin's haunting questions: where do we come from, what are we, and where are we going?

Liminal lives

IVF technologies evolved rapidly after Louise Brown's highly publicized birth in Manchester's Oldham General Hospital in 1978. Forty years later, upwards of six million babies had been born worldwide using IVF and related techniques.[3] These babies brought intense joy and fulfillment to many families, but from the standpoint of science and ethics the population that drew greater attention was the one left behind in the shadows: the ghostly army of unimplanted embryos whose numbers grew alongside those that were implanted and achieved human status as full-term babies.

The division of IVF embryos into those that are born and those that are not corresponds to the basic purpose of infertility treatment: to produce healthy pregnancies and eventual births. IVF takes a toll on the body of the mother-to-be, who is injected with hormones to stimulate her ovaries to produce multiple eggs during a single menstrual cycle. In turn, the production of multiple embryos from the retrieved eggs allows successive transfers into the womb until a pregnancy begins. Conducting this invasive procedure with a single embryo from a single egg would be too chancy, with a high probability of failure, but if enough embryos are made to increase the likelihood of success, then some are likely to be left over after a successful implantation. Indeed, IVF guideline revisions over the past decade or so have progressively reduced the number of embryos recommended for a single transfer, as the risks of multiple births became more apparent. A 2006 Canadian guideline recommended a graduated approach, with up to four embryos transferable in women over 39, although the authors acknowledged that high numbers of multiple births would not be desirable (Min et al. 2006). The 2015 guidelines of the European Society of Human Reproduction and Embryology recommend the transfer of only a single embryo at a time, while noting that decisions should be based on factors including embryo quality and female age. "Supernumerary embryos," the guidelines state, "may be cryopreserved, donated to research or discarded, according to their quality, patient wishes and national legislation" (ESHRE Guideline Group 2015, 19). The provisos at the end of that sentence indicate in the barest terms that more is at stake in the preservation of supernumerary (or spare) embryos in IVF labs than mere scientific judgment.

IVF embryos are subject to the wishes of both "patients," or those undergoing infertility treatment, and national legislatures – but those two sources of authority are not wholly independent. From the very earliest days of IVF, questions arose over how to classify the embryos awaiting transfer or no longer needed for fertility treatment (Jasanoff 1995, 170–1). Were they the property of the egg donor or the donor couple? What was the clinic's responsibility for the proper handling and disposal of these highly personal and sensitive materials that represent people's hopes for continuing their family lines? Did the state, as ultimate guardian of citizens' lives, have any stake in these proto-lives, capable of being maintained for years once embryo freezing became technologically feasible, and if so how should those interests be expressed and asserted? Answers to these questions began diverging across the Western world, and eventually globally, revealing different compromises among biomedical research, pro-natal policies, and respect for the liberty of reproductive choice. Repeatedly, the lab-made IVF embryo forced states, citizens, and scientific communities to confront the question "When does human life begin?" And, just as often, answers diverged across nations, along with decisions about who holds the ultimate authority to answer that question.

Inevitably, prior debates on the value of life and the state's responsibilities toward it colored determinations of the ethical and political status of spare IVF embryos. These, in turn, were rooted in different national treatments of religious and cultural differences with respect to reproduction and the family – that is, in divergent constitutional settlements with respect to emergent human life. Strikingly different regulatory pathways opened up in Britain, Germany, and the United States, despite their fundamentally equal commitment to protecting both lives and human values. All three are democratic societies with long traditions of political and humanistic thought; all three, too, are committed to free inquiry, the advancement of science, and liberal political traditions that demand publicly defensible grounds for state intervention, especially where intimate or private decisions are at stake. They share the same histories of science, with intertwined, often cross-cutting, developments in biology and biotechnology. Yet, national policies for the treatment of IVF embryos and derivatives such as hESCs radically parted company, underwritten by different ethical principles and institutional practices.

Britain, where Louise Brown's birth ushered in the IVF era, proved in some ways most permissive toward technologically assisted reproduction, while at the same time regulating it most comprehensively. It soon became clear that the repercussions from Brown's birth could not be contained within the walls of labs or clinics as issues to be decided by science alone. A new biological entity, the "test tube" embryo, had been introduced into the world. How should this entity be named and known, and who should be responsible for it? True to its standard operating practices in areas of grave policy uncertainty, the British government appointed an expert committee chaired by the Oxford moral philosopher Mary Warnock to advise the government on these questions and to restore moral order surrounding the revolutionary advances in infertility treatment and embryo research. Two years later, in 1984, the committee issued its now canonical report recommending, most importantly, that no research should be permitted on human embryos beyond fourteen days of development (Warnock 1984). Six years later still, Parliament created by law a special regulatory body, the Human Fertilisation and Embryology Authority (HFEA), to supervise all decisions involving the troubling liminal objects that Edwards and Steptoe had created.

The fourteen-day rule, as I have shown elsewhere, gained acceptance in Britain by appealing to well-established national traditions of building public reason, or "civic epistemologies" (Jasanoff 2005; Jasanoff and Metzler 2018). Judgment was entrusted in the first instance to that characteristically British institution: a "safe pair of hands" – in this case, Mary Warnock, who embodied not merely philosophical expertise and high academic standing, but also a sure intuition for how to meet public expectations through commonsense moral line-drawing. Warnock, later elevated to a peerage for her services to the nation, held an unswerving vision of what was required of her and her committee. It was to secure public acquiescence for a delicate balancing act. The committee's task was not, she insisted in a 2017 interview, to answer a factual question ("When does life begin?"), but rather to resolve a sensitive issue of classification in a moral gray zone: "A live embryo in the laboratory was a completely new object which had never existed before, and its moral status had to be discussed and clarified. Was it to be treated as a collection of cells or as a baby?" (quoted in Hurlbut et al. 2017, 1041).

The best response, Warnock and her committee members concluded, was to create an enforceable time limit on research that would appeal to public common sense, assuage any fears that an experiment "would possibly kill a fully formed baby," and yet leave scope for reasonable scientific inquiry (Hurlbut et al. 2017, 1041). Those limits should reflect the ideal of "a kind of society that we can, all of us, praise and admire, even if, in detail, we may individually wish that it were different" (Warnock 1984, 1). With advice from the developmental biologist Anne McLaren, the committee decided that an important moment in embryonic life comes around a fortnight after fertilization, when cells begin to specialize, the possibility of twinning ends, and the "primitive streak" or beginnings of the nervous system appear. From this point onward, the rapidly developing "collection of cells" looks more like a precursor to a full-fledged human, and hence no longer a fit object for research. This, then, offered a bright line that many might accept, whatever their personal views on when life begins. To underline that embryo research would never become a slippery slope, the Warnock report proposed making it a criminal offense to keep IVF embryos alive after fourteen days. Many years later, Warnock matter-of-factly recalled that she settled on the number fourteen more to satisfy society's need for clarity than out of any biological necessity: "I chose 14, rather than 13 or 15, simply because everyone can count up to 14; a fortnight is a good, memorable number, and records can be kept week by week" (Hurlbut et al. 2017, 1041). In other words, a morally significant inflection point in embryonic development was designed to align well with society's existing practices of time-keeping.

Similarly pragmatic sensibilities marked the committee's thoughts on administration. To implement the report, the committee recommended a single, central body, the HFEA, to license and monitor activities involving embryos. Successive HFEA chairpersons proved capable of speaking in the register of shared values and so carried on the Warnock report's strategy of consensus-building through appeals to common sense. Thus, the late Lisa Jardine, a noted Renaissance historian with no specific expertise in ethics or reproductive medicine, lightly dismissed even the possibility of serious disagreement on when life meaningfully begins. In a 2008 interview, soon after she was appointed chair of the HFEA, she told a *Guardian* reporter:

> St Augustine, who I greatly revere as a great father of the church, believed that the child became human when it kicked in the womb, so that would be 19 weeks. The other religions of the book have post-14-day beginnings of consciousness, so only 21st-century Catholicism has this problem, and there are thinking Catholics who I greatly respect who genuinely believe the Catholic church should roll that back. In which case, we'd all be able to talk to one another. (Boseley 2008)

There is here a sense of confident empiricism, backed at the same time by tolerance toward religious belief, a presumption that if people would only "think," they would come together around a sensible common norm, one ratified by centuries of religious thought reaching all the way back to St. Augustine. That sentiment echoes more than thirty years later the Warnock report's faith that it is possible to have the kind of society that all members can "praise and admire," even when differing on details here and there. And one indispensable element in forging that consensus is the voice of a Warnock or a Jardine who, through some combination of professional experience, social standing, and record of public service, can presume to speak sense for the British public, if not for all humanity.

It would be hard to imagine a constitutional order less like Britain's than the political struggle over governing emergent forms of human life that erupted in the United States. At bottom lies that nation's peculiarly toxic politics of abortion, unresolved since the Supreme Court's landmark 1973 decision in *Roe v. Wade*. There, the Court recognized women's basic right to terminate their pregnancies, although the right has been overlaid with an increasingly incoherent and restrictive patchwork of state-by-state regulation. Those dynamics polarized long since into sloganeering, pitting the "pro-choice" movement against its "pro-life" adversaries, the former backing adult women's right to choose their own reproductive futures, the latter bent on staking out personhood claims, and state intervention, on behalf of the unborn. Although Mary Warnock's question (is the lab-created embryo "to be treated as a collection of cells or as a baby"?) resonated in the US context, American answers were bound up from the start with preconceived notions of fixed and inalienable rights triggered by the factual question the Warnock committee so assiduously avoided: when does life begin? Public opinion concerning the beginning of fetal life and the moral worth of the IVF embryo thus remains sharply divided, and settlements are achieved not through elite leadership and guided consensus, but at the polling booth and in the shadowy dynamics of federal legislative politics.

Several episodes over approximately two decades of political strife illustrate why the constitutional pushes and pulls of the US system have failed to stabilize a governance paradigm similar to Britain's fourteen-day rule and oversight through the HFEA. The first occurred in 1996, with the passage of the Dickey–Wicker amendment to the federal appropriations bill passed in January of that year. Authored by Republican congressmen James Dickey of Arkansas and Roger Wicker of Mississippi, the amendment simply bans the use of federal funds to create a human embryo for research or to support "research in which a human embryo or embryos are destroyed, discarded, or knowingly subjected to risk of injury or death greater than that allowed for research on fetuses in utero."[4] The amendment, annually renewed since its first enactment, was never publicly debated: a crisis atmosphere following two paralyzing government shutdowns made all sides hastily come together on getting an appropriations bill approved as quickly as possible. In that setting, a simple partisan stroke of the legislative pen overrode both scientific expertise and collective moral judgment to define what would count as viable human life at bioengineering's moving frontier.

Another key episode shifted the site of biopolitics from Congress to the White House, illustrating how presidential power can steer bioethical judgment in a decentralized and pluralistic system. Here, the triggering events grew out of James Thomson's historic success in 1998 in deriving hESCs from spare IVF embryos, a move quickly emulated and productively extended in labs throughout the United States and the world. Although Dickey–Wicker forbade federal funding of research on embryos, Harriet Rabb, lead lawyer for the NIH, interpreted the ban as not applying to subsequent research on embryo derivatives such as hESCs. The Clinton administration accepted Rabb's reading, creating in effect a bifurcated funding scheme in which private funds could be used to derive stem cells, but federal funds could only be applied to downstream research using already existing cell lines. George W. Bush's surprise election to the presidency in 2000 upset that pragmatic balance. Under pressure from the religious wing of his party to halt the destruction of frozen embryos, Bush announced that his administration would not allow the ends of research, however valuable, to take precedence over the means. At his first public address to the nation on August 9, 2001, the new president declared that NIH funds would be used for research using only the sixty or so privately funded cell lines already created by that date, "where the life and death decision has already been made."[5] Stem cell scientists were forced to work with and around this restrictive policy, catering to minority religious sentiments, until Barack Obama's election in 2008 turned the wheel back in their favor. Lifting restrictions on NIH funding for hESC research, President Obama appealed to a different moral imperative, America's pioneering role in research: "Today, using every resource at our disposal, with renewed determination to lead the world in the discoveries of this new century, we rededicate ourselves to this work."[6] To be sure, that work would be undertaken with respect for life. NIH guidelines provided for a new layer of institutional ethics review to ensure that research would be done only on embryos secured with the donor's consent and without violating bans on human cloning or other morally transgressive boundaries such as creating human–animal chimeras. Both Bush and Obama, in short, established a controlling, nationwide narrative on the limits of experiments with life, settling through executive judgment the moral question of what respect is due to lab-engineered entities straddling the boundary between life and nonlife.

Yet a third dramatic chapter in stem cell politics opened in the courts. James Sherley, a stem cell biologist once employed at MIT, and his co-plaintiff Theresa Deisher sued the federal government to block the NIH from funding research under what they claimed was a misinterpretation of the Dickey–Wicker rider. They won a preliminary injunction from Judge Royce C. Lamberth of the District Court for Washington, DC, who concluded, to the consternation of stem cell biologists, that the ban on embryo-destroying research must apply not merely to the initial derivation of the stem cells, but to all subsequent research involving these entities. The NIH's line-drawing, Lamberth held, made no sense: "Simply because ESC research involves multiple steps does not mean that each step is a separate 'piece of research' that may be federally funded, provided the step does not result in the destruction of an embryo."[7]

It fell to the higher appellate courts to step into fray, and they did so by invoking the traditional separation of powers within the federal government. The Court of Appeals for the DC Circuit ruled in 2012 that Congress had left ambiguous what counts as "research in which a human embryo or embryos are destroyed." According to well-established prior law, an executive agency such as the NIH is entitled to adopt an authoritative reading of the law where the legislature has not made its meaning clear. Moreover, the court concluded, Congress had tacitly accepted the NIH's reading of the Dickey–Wicker amendment by renewing it unchanged year after year. Hence, the injunction was not warranted and research could continue in accordance with the Obama administration's guidelines. The Supreme Court let the appellate decision stand, and US stem cell researchers heaved a collective sigh of relief, preferring to live with the relatively friendly ethical supervision demanded by the NIH than with possibly more radical uncertainties introduced by Congress and the courts.

Germany's path through this murky territory followed an entirely different course, except that here too the ultimate meanings given to embryos and hESCs involved complex negotiations between science and medicine on the one hand and law and politics on the other. In Germany, concern for human life asserted itself through a postwar constitutional order in which a state that itself needed to be reborn and reintegrated into the community of nations formally dedicated itself to protecting human dignity. Much of Germany's contemporary constitutional politics can be seen as a politics of penance and redemption, an effort to ensure that no state-sponsored project of mass slaughter would ever again be contemplated, much less carried out. To this end, Article 1 of Germany's Basic Law states as a moral absolute, "Human dignity shall be inviolable [*unantastbar*]. To respect and protect it shall be the duty of all state authority."[8] Article 2, Section 2 elaborates on that first principle by declaring: "Everyone shall have the right to life and physical integrity." These are of course abstractions. They do not answer in and of themselves

108

who counts as a person, when the right to life attaches to such beings, or precisely what it means for the state to respect or protect human dignity.

As in the United States, Germany was to some extent primed for the moral challenge posed by IVF embryos through prior engagements with abortion. German reunification in the early 1990s required two formerly distinct states – East and West Germany – to merge their constitutional thinking into a single, integrated whole. Abortion laws on the two sides proved especially hard to navigate (Jasanoff 2005, 159–61). East Germany had permitted abortion virtually on demand in the first trimester of a pregnancy, a scheme the West rejected because it granted no moral status to the developing fetus. In the West, abortion at that time was allowable if women could show they met certain medical or social indications. The compromise hammered out with some difficulty established a firm biological beginning for human life, namely, at the moment when the nuclei of sperm and egg fuse. From that moment onward, German law holds, human life exists in a form that the state is committed to preserving and protecting consistently with its other obligations toward citizens' lives. Abortion in reunified Germany remains a criminal offense, except when carried out under medical supervision; in practice, however, medical approval is relatively easily available, especially for first trimester abortions. This division of responsibility between the state and the medical profession has allowed Germany to maintain a balance between explicit public condemnation of the destruction of emerging life and a relatively humane approach to the treatment of women facing the burdens of unwanted pregnancy and childbirth.

This constitutional settlement left no room, as in Britain, for a fourteen-day rule, in which the early embryo is severed in effect from the rest of the human community as belonging more to the category of "cells" than of "babies." Nor does German law allow for the kind of case-by-case legal challenge and resolution that surrounds abortion and, to some degree, stem cell research in the United States. Most clearly, if human life begins with fertilization, then the very idea of creating such lives for research, or discarding them as if they are mere waste, is anathema to the German legal imagination. How then to ensure that infertility treatment, considered an unalloyed social good, would continue, with access to the latest technological breakthroughs, but without sacrificing the incipient human lives created in the lab to serve this beneficial purpose? To resolve this problem, Germany enacted one of the most restrictive regimes of ethical and legal oversight over IVF in all of Europe.

The surest way to ensure that human lives will not be created only to be sacrificed is not to produce such lives at all. The German Embryo Protection Act, passed in 1990 and later amended to accommodate stem cell research and prenatal genetic diagnosis, attempts to do just that. The law makes it a criminal offense to fertilize human egg cells for any purpose other than to start a pregnancy, as well as to perform other transgressive activities such as cloning human embryos or making human–animal hybrids and chimeras. Since only as many embryos are formed as may be safely implanted, German infertility practice has not produced the vast quantities of unused embryos that have been generated in the more permissive and less tightly regulated

US IVF clinics. Against this background, Germany in 2002 debated whether to permit stem cell derivation and research, and reached its own somewhat split and incoherent conclusion.

The German Bundestag, like the British Parliament and the US Congress, had to steer between competing moral sensibilities, but in Germany these were more formalized and more clearly associated with major political party positions than in the other two countries. All sides agreed that the Embryo Protection Act simply forbids the derivation of stem cells from fertilized embryos that are entitled to respect as forms of human life. But could German scientists import stem cell lines from abroad, without officially participating in the destruction of embryos? Opinion divided. On one side was a report from an ethics committee appointed by Chancellor Gerhard Schröder that recommended imports by a relatively narrow margin; on the other side was a report from the German Parliament's own inquiry commission recommending against imports by a somewhat larger margin. A stylized debate in the Bundestag divided along predictable party lines: the conservative Christian Democrats categorically opposed stem cell imports; the free market Liberals advocated loosening all controls on stem cell derivation; and the ruling Social Democrats opted for imports while speaking out strongly against research that results in wastage of embryos (Jasanoff 2005, 197–8). The Social Democratic position against derivation but for import won the day by a wide margin, and still remains the law.

In the end, the German stem cell regime accepted a moral contradiction no less striking than that in America, but the compromise took a different form.

In both countries, the state refused official support for destroying *in vitro* embryos by using them for stem cell derivation, although the sanction in the United States was denial of public research funds, whereas in Germany it was the threat of criminal prosecution. Yet, both states continued to permit research using stem cells, thus turning a legally blind eye to their derivation – so long as that step did not represent state policy. In the United States, a bifurcated public–private system of research funding allows private foundations and donors to sponsor research centers such as James Thomson's institute at the University of Wisconsin. In Germany, importing stem cells from abroad absolves the nation and its scientific community of complicity in destroying potential human lives, but it is a workaround, a bit of legal jiujitsu that satisfies the nation's need for legal clarity regarding when life begins without significantly burdening science. Only Britain solved the problem without apparent contradiction, by declaring the early embryo not to be the same category of thing as the post-fourteen-day construct, but that rule too proved vulnerable to biology's demand for greater control over the classification of living things.

Ambiguous achievements

Criticized on both scientific and philosophical grounds, as lacking firm foundations in logic or biological fact, the Warnock committee's fourteen-day rule nonetheless won wide acceptance as a workable compromise. For all practical purposes, it became the global standard in embryo research. As bioengineering moves forward, however, and technologies develop for keeping embryos growing *in vitro* for longer periods of time, there have been increasing calls from within the scientific community to revisit the rule and see if it still makes sense as a matter of global public policy (Weintraub 2017). Advocates of abandoning the rule dismiss it as driven by expediency rather than principle, and hence of questionable moral significance. A new era, with its new technological possibilities, this argument implies, is ripe for a different compromise, one that would lengthen the amount of time available for research on the developing embryo, since after all – just as Hans Jonas claimed at the 1976 Airlie House conference on biomedicine and society – it is technology that now supplies the most important questions, objectives, and limits for science.

These views were aired in a 2017 review in *Nature Biotechnology* of opinion concerning the fourteen-day rule; policy scholar Aaron D. Levine reiterated the rule's lack of foundational support, asserting that it "represented a compromise between competing views more than an elegant and convincing philosophical argument" (Hurlbut et al. 2017, 1032). Prominent scientists, such as the embryologist Robin Lovell-Badge and reproductive biologist Alison Murdoch, saw no immediate technological advances that would allow normal development of *in vitro* embryos beyond the fourteen-day limit. Lovell-Badge, for example, noted: "There are probably several limitations of the current methods that are likely to compromise normal development much beyond 13 or 14 days." But the scientists' skepticism leaves open the possibility that it might be appropriately revisited if technology eventually enables what is currently impossible. Indeed, Murdoch herself implied as much in tying the fourteen-day rule to a now obsolete state of knowledge:

> The Warnock solution to "When does life begin?" was a consensus decision that led to the unscientific but legally precise definition of an embryo: the act of mixing an egg with sperm. The 14-day rule never intended to be a moral line; it was a compromise, a pragmatic decision based loosely on knowledge of embryo development in 1990. (Hurlbut et al. 2017, 1033)

These observations, however, all missed the point of Warnock's philosophical position, one she asserted with fierce clarity well into her nineties. The "Warnock solution" was not, as Murdoch asserted, a pragmatic answer to the question concerning when life begins, although her confusion was not hers alone. Indeed, as Warnock herself took pains to point out in her 2017 interview, she never "succeeded altogether in getting members of the committee, let alone members of Parliament, to grasp that *this was not a question of fact*" (Hurlbut et al. 2017, 1041; my emphasis). What Warnock saw more clearly than others was that biotechnology had produced an ambiguous new entity, the "live human embryo outside the uterus," and its moral status had to be determined for society through an authorized form of deliberation. This was at bottom a moral and even a legislative question, one for society to decide, and her committee laid the groundwork by forging a compromise, as legislators must do even in areas of contested moral judgment.

This was a mode of sense-making that could not be left to science alone, because science has nothing definitive to say about how society should take up and deal with its boundary-straddling creations. To criticize the

Warnock committee's consensus-building process, then, one would have to turn not to science but to principles of democratic theory – for example, questions of inclusiveness and adequate deliberation. Even on this front, one may note that the committee's conclusions were eventually ratified by Parliament, providing the ultimate stamp of democratic legitimation. The point to reaffirm is that the committee produced not an "unscientific" definition of an embryo but, rather, a recognizable, commonsensical stopping point for research on borderline human life, a limit beyond which science should not go in a society that all could "praise and admire."

The tortured ethical, legal, and political history of the IVF embryo and the fourteen-day rule illustrates, above all, how making sense of life in the gray zones of biotechnology involves struggles for authority between a progress-hungry, transnational science and situated social demands for ethical reflection and public engagement in the governance of life. Scientists, even at their most restrained in assessing what is possible, tend to veer toward deterministic positions that equate what can be done with what should be done. Against that backdrop, we might keep in mind J. Benjamin Hurlbut's reminder from the same 2017 review:

> that it [the 14-day rule] *was not contingent on the state of the science*, but was grounded (at least in the UK) in *public moral judgment* – and in accounts of how democratic societies ought to go about making such judgments. Thus revision ought not be taken lightly. At stake is society's trust in scientists to abide by ethical limits and to treat public ethical concerns as serious and significant and not as a mere political problem. (Hurlbut et al. 2017, 1035)

We turn in the next chapter to another strategy by which science has sought to maintain control over its own productions, so as to guard against what many see as premature and excessive populist intervention into the pace and direction of research. This is the use of language to claim entire research territories as governable by science alone, because – unlike the human embryo – the entities within those territories lack moral significance. Their sole function is to advance scientific inquiry, and hence their use and fate are only of concern to science. These language games are integral to the repertoire of devices through which the life sciences assert their power not merely over what life is, but also over the purposes toward which we should direct the manipulation and redesign of life.

5

Language Games

Without communication there could be no shared meaning. Humans in particular have the gift of language. We make sense of things by naming them, finding words to describe them, and speaking with others to make sure our understanding of things matches that of those in our communities. In the Book of Genesis, God the creator offered Adam a kind of junior partnership by allowing him to name the other entities in the Garden of Eden, holding back "to see what he would call them: and whatsoever Adam called every living creature, that was the name thereof."[1] And so the biblical Adam named the birds and the animals, and in due course even his female counterpart Eve, and those became the names by which all were known. But language as we use it in communication consists of much more than names, or dry dictionary entries with their definitions. Language shapes our thinking, makes things come to light or disappear from view, and participates in the channeling of power – possibly nowhere more so in the modern world than when language designates some things as scientific, and hence a part of what we take to be immutably natural.

Philosophers of knowledge have long been interested in the nature and power of language. In Britain, Ludwig Wittgenstein, the giant of twentieth-century analytic philosophy, coined the term language game to demonstrate that words are more than simply what we say. They do their bit to organize our actions, as when a person goes into a grocery store and asks for five red apples and the shopkeeper has to find the apples, sort them by color, and count to five to fill the order (Wittgenstein 1953, 2–3). Those actions, moreover, are inseparable from the meaning of the words themselves. A word like five means something, Wittgenstein argued, only because it is used in a certain way, so that word and action cannot be kept distinct. For this reason, Wittgenstein concluded, "to imagine a language means to imagine a form of life" (1953, 8).

Historians and sociologists of science have approached the indisputable connection between word and action in a variety of ways, but most agree with Wittgenstein that to decide how to speak about something is also to determine how to navigate with and around the things that language designates. The very rise of experimentation as a form of life within science demanded the formation of what the historians Steven Shapin and Simon Schaffer (1985, 61–9) called a new "literary technology." Robert Boyle, a leading actor in that story, self-consciously set about this work of craftsmanship, through a tract on "experimental essays" that advocated for a particular style of writing. "[P]lain, ascetic, unadorned (yet convoluted)," that style became identified with the rise of a modest and impersonal science in seventeenth-century England, a new experimental form of life. Since those early, heady days of the scientific revolution, specialization has splintered the language of science into many esoteric dialects that, in effect, signal membership within what Ludwik Fleck, the Polish physician turned sociologist of knowledge, called a thought community (*Denkgemeinschaft*). Such a community "becomes isolated formally, but also absolutely bonded together through statutory and customary arrangements, sometimes a separate language, or at least special terminology" (Fleck 1981, 103). Coincidentally, these specialist and expert languages not only define and isolate an in-group but also guard the gates against would-be interlopers, much as the ancient Hebrew word *shibboleth* was used as a password to distinguish between those inside the Israelite speech community from foreigners who could not correctly pronounce the first consonant.

The concept of discourse elaborated by Michel Foucault is related to language games and forms of life in that it also calls attention to ways in which words and styles of speaking relate to features of the context in which they are used. Foucault's primary concern with the human sciences makes his views especially relevant to understanding the constitutive role of discourse in relation to life in its varied aspects. Discourses that we recognize as, say, medicine or law, or economics, psychology, or bioethics, do not owe their force to words alone, but also to the background beliefs and practices that authorize particular kinds of statements to be made about particular sorts of topics to particular listening, and comprehending, audiences. These forms of authorization – defining who can speak in a given discourse, to whom and about what, and from which institutional positions – make it impossible to split off power from knowledge (Foucault 1969, 50–5). As Foucault himself put it: "In fact, every point in the exercise of power is a site where knowledge is formed. And conversely every established piece of knowledge permits and assures the exercise of power" (Morris and Patton 1979, 62). Both knowledge and power are asserted through language, but context may give the same words different effect. "Off with his head" meant, literally, death by decapitation when spoken by a monarch in Shakespeare's England; by Lewis Carroll's time, the Queen of Hearts only came across as a blind and ineffectual fury when she went around muttering the same injunction.

Foucault's monumental body of work repeatedly turns to the constitutive role of discourse, both generally in its shaping of our knowledge and self-understanding and specifically in its applications to criminality, madness, sexuality, and other human classifications. The objects of discourse, those things that are brought into view through words, are not, he insisted, simply out there to be found: it is not "enough for us to open our eyes, to pay attention, or to be aware, for new objects suddenly to light up and emerge out of the ground" (Foucault 1969, 44–5). Rather, those objects only exist in consequence of the prevailing discursive practices of the day – practices that provide the conditions of possibility for naming and identifying those kinds of objects in the first place. Correspondingly, powerful ways of making sense of lives and worlds have the effect of making some things *not* apparent, not emerge out of the ground, although they might have been recognizable in a different discursive context. Professional discourses, moreover, will normalize some things, making them appear banal, whereas others are picked out as deviant and subjected to sanction or correction by society's ordering forces.

Language Games

Which discourse governs becomes particularly important in decisions about what to do with novel entities,
objects that like the IVF embryo discussed in the preceding chapter would not have existed without human
manipulation of life. It matters especially whether lay
naming practices and meanings will come into play, or
whether science can claim the objects it has made for
itself. Below, we look at key moments when the modern
life sciences have struggled for control with other discourses to decide whose ways of describing reality, or
which language games, will apply, and with what consequences for making sense of life.

Biotechnology: pure or entangled?

Origin stories matter in claiming control over the definition of life. How does biotechnology narrate its own? As
human ability to manipulate the material components of
life has advanced, those wielding the new powers have
had to confront questions about what it is they are
doing. Is biotechnology just another branch of science,
with its claims to freedom and autonomy, because it is
the work of the human spirit in its ceaseless quest for
knowledge? Or is it, as Hans Jonas pointed out in 1976,
a shift in the very nature of science, led by the lure of
technological applications and leading in turn to novel
demands for accountability? If, as Jonas insisted, the
program of the life sciences is "increasingly defined by
extraneous interests rather than its own internal logic,"
then science no longer has recourse to "the alibi of
pure, disinterested theory" and must instead submit to
judgment for its social actions (cited in Steinfels 1976,

22–3). Tensions over naming and control have continually resurfaced in the history of modern biotechnology. On the one hand are discourses of purity, often told by a collection of interests, comprising scientists, start-ups, and the biotech industry, who seek to connect their work to nature and to disinterested science. On the other are discourses of embedding and application, more often told by philosophers and social critics who wish to make visible science's thick ties to commerce and utility, and to corresponding moral visions – thereby reinforcing the case for expanded social oversight.

The purity story can be found in the timelines of biotechnology that have circulated for decades in industry-oriented websites on the internet. Some are committed to establishing that human beings have been altering nature for centuries, if not millennia, and hence imply that there is nothing new in the kinds of work moving forward in today's advanced laboratories. Others stress a continuity with the history of science, citing connections between today's discoveries and other notable achievements in biomedicine. A site created by a leading trade association, the Biotechnology Innovation Organization (until 2016 the Biotechnology Industry Organization – a name change that itself speaks volumes), puts the start date for biotechnology at 500 BC, when an antibiotic derived from moldy soybeans was put to medical use; from about AD 100 the timeline fast-forwards to 1761, when Edward Jenner pioneered vaccination for smallpox.[2] That trajectory foregrounds biotechnology's links to medical discovery, as does a site maintained by the Biotechnology Institute, another industry organization, which equates the birth of biotechnology with Robert Hooke's first description of a cell in 1663.[3] Most circumspect is the timeline maintained by Amgen, an early but still influential biotech company, which starts the clock running in the 1950s with the establishment of the HeLa cell line and the discovery of the structure of DNA, although a page devoted to the history of biotechnology goes back to the Hungarian Karl Ereky's first use of that term in 1919 (see also Bud 1993, 32–6).[4] The noteworthy point is that it is commercial biotechnology that most assiduously projects a pure, unbroken lineage from the earliest biomedical discoveries to today's profitable technological enterprises.

A very different narrative emerged in Germany in the 1980s, when the German Parliament began debating how to regulate biotechnology. The Green Party, elected to the Bundestag for the first time in 1983, made this one of its prominent causes. In 1986, the Greens adopted a position paper that, like Hans Jonas a decade earlier, stressed biotechnology's rupture from a past of pure science and free inquiry. It was a branch of applied knowledge, the Greens asserted, oriented toward use rather than discovery, and further tainted by industry support and cozy relations between government and industry. These entanglements posed potentially catastrophic risks for humanity, the Greens further argued, especially for vulnerable populations in the Third Word, and hence called for a suspension of the constitutional protection normally afforded to free scientific inquiry. The critics demanded more transparency and broad public participation in decisions involving biotechnology, as well as a full consideration of alternatives each time a genetic engineering experiment was put forward (Jasanoff 2005, 64–6). The 1990 German law regulating genetic engineering did not accept all of the Greens' sweeping requests, but it did acknowledge the need for targeted legislation on biotechnology, something that the US government never felt to be necessary.

While Germany struggled to craft a political compromise between its liberal and left constituencies, US biopolitics too demonstrated the power of discourse by attempting to cleanse itself of a label that increasingly struck both industry and government as inaccurate, possibly scary, and overly general. Biotechnology companies, the *Los Angeles Times* reported in 1986, "can't agree whether they want to be identified as part of an industry or simply the makers of tools for a variety of other industries" (Walters 1986). Definitions, these companies recognized, might make a difference to regulatory sensibilities. Indeed, some argued at an industry conference that the encompassing term biotechnology invited a wary public to ask for sweeping controls, when what was needed was a more precise and differentiated approach to particular products. Henry I. Miller, special assistant to Frank E. Young, commissioner of the Food and Drug Administration (FDA), called the situation "biotechnology creep," and sought to resist it.

Within the US government, too, Robert Bud describes an interagency struggle over terminology and meaning that turned "Kafkaesque" and "resulted in the polarization and politicization of the very word 'biotechnology'" (1993, 212). Normally invisible bureaucrats turned into bare-knuckled street-fighters, taking to the news media to express their views. Among the most vocal were Miller and Young of the FDA. Together, they wrote an opinion piece in the *Wall Street Journal* in January 1987 seeking to recapture the escaped genie of language and put it back into bottles that expert regulators could comfortably control:

Language Games

Defining the terms "biotechnology" and "genetic engineering" isn't an easy task, since the terms don't represent natural groupings of processes or products. They connote something different to individual commentators, journalists, organizations, congressional staffers and members of the public. The terms are ambiguous, the source of much confusion and little advantage, and we would do well to return to more specific and descriptive terms. Laymen would understand them better and the complaints of poorly informed critics would be put into perspective. (Miller and Young 1987)

Instead of biotechnology, the authors proposed terms that would be more "informative, lucid and useful," such as biochemistry, microbiology, ecology, recombinant DNA, live vaccines, therapeutic monoclonal antibodies, and enzymes. The challenge they saw was to replace a word in widespread public use with words proper to science and scientists – words that laypeople would not be able to hijack or tie to their agendas of political control. Biotechnology was a term anyone could appropriate; monoclonal antibodies and enzymes were less amenable. This reassertion of expert authority, or scientific power grab, was proposed in the name of conveying meaning more clearly, countering lay confusion with technical exactitude, and deterring demands for oversight by an ill-informed public.

The embryo: what's in a name?

The same theme of using the alleged precision of scientific discourse to avoid lay confusion surfaced in connection with the *in vitro* human embryo, that highly contested object occupying a gray zone at the nexus of life and nonlife, or science and politics. Here too, however, the desire for precision masked a struggle for control. From the moment the Warnock committee split embryonic life into pre- and post-fourteen-day phases – the former an open door to research, the latter triggering moral concern – the *in vitro* embryo generated fierce debate about the meaning of human-ness and the limits that should be imposed on research. Language in this contested environment simply could not be neutral. Names would position objects on one side or the other of a regulatory divide with profound moral and political consequences. Applying the term "pre-embryo" to the pre-fourteen-day *in vitro* construct, for example, implied to some that a distinct, or lesser, biological entity existed before the stipulated cut-off date when the cells became full-fledged embryos. In keeping with its advisory role, and mindful of Parliament's needs, the Warnock report itself carefully avoided using this charged term and instead simply opted for a temporal limit on allowable research. Yet, few seemed able to wrap their heads around the fact that the fourteen-day rule was not about defining an ontology – a new kind of being – so much as it was a rule of conduct showing respect for the developing human, even if created in a glass dish in a laboratory.

Debates in and out of legislative chambers over the next few years testified to the wisdom of the Warnock committee's choice, as critics charged that the term pre-embryo was merely a linguistic ploy used by science to hide its instrumentalizing of human life. Erwin Chargaff, eminent Hungarian-born biochemist at Columbia University, who had fled the Nazis, was especially passionate on the subject. The pre-embryo, he wrote in 1987, "is a designation that appears to me entirely unjustified. I fear that it has merely an alibi function." That function was to produce a justification for research that Chargaff, a persistent critic of bioengineering, found deeply problematic:

> Had I anything to say I would certainly proscribe the production of human embryos for experimental purposes. Scientific curiosity is not an unbounded good, although I would not deny that, for instance, the problem of cell differentiation during embryonic growth is of great interest. Restraint in asking the necessary questions is one of the sacrifices that even the scientist ought to be willing to make to human dignity (1987, 199–200).

Earlier he wrote, in evident despair: "My generation, or perhaps the one preceding mine, has been the first to engage, under the leadership of the exact sciences, in a destructive colonial warfare against nature. The future will curse us for it" (1976, 940).

In Britain, too, voices of common sense spoke out against the "alibi function" served by the term pre-embryo. In the House of Commons, Ken Hargreaves, a leader of the anti-Warnock forces seeking to ban all embryo research, condemned scientists "for inventing the 'Humpty Dumpty word "pre-embryo"' to diminish the status of human individuals at the earliest stage of development" (Mulkay 1997, 30). In a similar vein Lord Kennet, speaking in a 1988 House of Lords debate, called it "purpose-serving language," and noted: "We do not call any other stage of the human life story something beginning with 'pre'. It is a negative definition which merely says that it is not an embryo in order to avoid the stigma of destroying embryos."[5] The 1990 Human Fertilisation and Embryology Act that resulted from six years of vigorous debate avoided the term and all its connotations, stating only that "embryo means a live human embryo where fertilisation is complete" and includes "an egg in the process of fertilisation."[6]

Science and technology move on, and language games move along with them. As we saw in Chapter 4, research has advanced to the point where embryos can potentially survive *in vitro* for longer than fourteen days, thereby allowing researchers to peer inside what has hitherto been a *terra incognita* of developmental biology. Robin Lovell-Badge, of London's Francis Crick Institute, noted in a 2017 interview with *Nature Biotechnology*: "We know almost nothing about human embryo development in the period (sometimes referred to as the black box period of human development) between 7 days, when implantation occurs, and about 28 days, which is the earliest stage at which it is possible to obtain embryos after terminations (abortions)" (Hurlbut et al. 2017, 1031). The temptation to probe into that unknown zone by scrapping the fourteen-day rule grew stronger for some researchers, such as the Rockefeller University's Ali Brivanlou, whose lab (along with that of Magdalena Zernicka-Goetz at the University of Cambridge) had discovered how to keep embryos alive for at least fourteen days. For scientists working on these frontiers, technical feasibility was enough to warrant liberalizing the rule. Echoing Gauguin's Tahitian masterpiece, but with a scientist's confidence that his methods would yield the definitive answers to questions of life's origin and purpose, Brivanlou told a reporter from National Public Radio: "If I can provide a glimpse of, 'Where did we come from? What happened to us, for us to get here?' I think that, to me, is a strong enough rationale to continue pushing this" (Stein 2017).

Today, too, we stand at another threshold of bio-technological possibility when techniques of synthetic biology have made it imaginable that self-organizing colonies of cells in a dish may take on the properties of developing human life. Once again, debates about how far science should go in pursuing these new lines of inquiry have crystallized into debates on what to call the entities that might in some sense be considered proto-human – as if the right choice of language will solve society's moral dilemmas. Once again, science's urge for exactitude has led to a claim of supremacy, suggesting that the correct terminology, whose descriptive adequacy only science can vouch for, will resolve any downstream moral problems. Yet, when science enters the game of linguistic innovation, as opposed to making new physical constructs, it never operates in unoccupied space, and must struggle for primacy with others' words and meanings which might call upon science to make its own accommodations. Naming new things that emerge within already rich and socially significant fields of meaning turns out to be extremely contentious business.

One such controversy erupted at a 2016 meeting held at Harvard University to discuss the future of the fourteen-day rule. Pressures to rethink the Warnock compromise were mounting from more than one side. Not only was technology seemingly at hand to prolong embryonic life in a dish beyond fourteen days, the project of scientists like Zernicka-Goetz and Brivanlou, but the possibility that cell masses might be produced with some embryo-like characteristics was also in the offing, as reported by Harvard biologist George Church and other scientists in his group. But what to call these organisms, or organoids, that were as yet developing in their own kind of gray zone, awaiting the bioengineer's ingenuity to make them "mimic" embryos? The terms favored at first were "self-organizing embryo-like structures (SOELS) and synthetic human embryo-like entities (SHELEs), which resemble natural human embryos in some ways" (Powell 2016).

The abbreviation SHELE, pronounced the same way as the name "Sheila," drew astonished responses from the Harvard audience, with participants noting the incongruity of giving a familiar woman's name to a proto-human entity that would also serve as a research subject. Adding a surreal cast to the debate, and despite the organizers' energetic efforts to secure diversity, all of the nine speakers at the event were men, though all were discussing rules for research with an object intimately tied to women's bodies, fertility, and self-awareness. The science writer Lauren M. Whaley (2017) commented on the "surprisingly one sided perspective" of this all-male panel, or "manel," when several qualified women experts in the room could have spoken to a number of the issues the meeting raised.

Public laughter and calling out had a marked effect on the game of language in this instance. A paper from the Church lab in early 2017 confected another name for the borderline objects emerging from their work: "synthetic human entities with embryo-like features" or SHEEFs (Aach et al. 2017).[7] These things, the authors concluded, might develop properties of moral concern, such as a capacity to feel pain, without progressing through the classic phases of embryonic development, such as the appearance of the primitive streak or proto-nervous system. They proposed that, instead of simply changing the fourteen-day rule to some other number, scientists, ethicists, and others should engage in a more foundational debate on which aspects of early human development are likely to arouse moral concerns – and to tie research limits directly to those factors. The term SHEEF, at any rate, began a quick migration into the mass media, picked up by publications such as the *New York Times* and *Scientific American* (Regalado 2017; Weintraub 2017; Zimmer 2017). Whether it would displace or compete for space with SHELEs remained for the time being an open question, especially as scientific speech could not easily be standardized across international borders and the United States was not the only location where relevant research was taking place. The entire terminological tempest remained enclosed within the circles of developmental biology and, to some degree, bioethics. It displayed, however, the zeal of science (and its not entirely representative "manels") to control the meaning of new forms of life through the Adamic power to name, and so to determine what a thing is and how it should be known.

Setting the terms of debate

The controversy over what to call self-organizing life in a dish – SHELEs, SHEEFs, or synthetic embryos – represents just one move, albeit a consequential one, in a much broader set of struggles about the right discourse for public debates on biotechnology. This is most definitely not neutral or value-free territory, as the choice of how to speak affects both the hierarchical relations among competing speakers and the relative visibility, or invisibility, of the objects under scrutiny. Scientists, by virtue of their frontline role in tinkering with life, have claimed primacy not only in naming their work products but also in setting the terms in which it is appropriate to debate them. In this section, we consider four episodes that left their mark on how language games shape deliberation and governance at the frontiers of emerging biotechnologies. These stories illustrate four overlapping patterns of deference through which science wrests from a compliant democracy the right to frame and shape the terms of debate: discursive narrowing, ethics lag, unruly publics, and inevitability.

In a much-quoted editorial in *BioScience*, Robert Sinsheimer, among the first scientists to acknowledge the power of recombinant DNA, mused about the future of a humanity poised to remake itself:

> With the advent of synthetic biology we leave the security of that web of natural evolution that, blindly and strangely, bore us and all of our fellow creatures. With each step we will be increasingly on our own. The invention and introduction of new self-reproducing, living forms may well be irreversible. How do we prevent grievous missteps, inherently untraceable? (1976, 599)

One answer, possibly not the one Sinsheimer would initially have advocated, has been to narrow the imagination of possible futures to those that scientists themselves can foresee, and hence in some sense also control, without fear of too much outside interference. As the historian Susan Wright noted in her comparison of US and British policymaking on biotechnology:

"An essential dimension of this early process was the progressive restriction of discourse surrounding genetic engineering – the formation of informal (often unspoken) rules about what was to be addressed, what was central and what was peripheral for decision making, and the norms of truth and falsehood" (1994, 114).

The 1975 Asilomar meeting, as we saw in earlier chapters, marked an important milestone in this process of discursive line drawing. Biologists started off down the path of genetic engineering with high enthusiasm, but also with genuine concerns about unintended consequences to health and the environment. Yet, money beckoned along with the lure of a new frontier, and Asilomar helped quell the biologists' uneasiness so as to make research seem, again, safe, normal, and morally defensible. The philosopher Michael Ruse describes the dramatic turnaround as a once-raucous debate abruptly fell silent: "Today in fact, much recombinant DNA work is allowed to go on virtually anywhere, so long as one closes the door when one is working and puts on a lab coat if the head of the team insists" (1981, 159–60). The Asilomar meeting produced an effective vocabulary of containment, and a corresponding set of work practices that reassured scientists and regulators that research posed no threats to health and safety. Publics were mostly kept at bay while decisions were reached in expert committees dominated by the very disciplines whose work was under scrutiny. Worries such as Chargaff's about breaching profound moral boundaries or Sinsheimer's about humans taking control of natural evolution – not to mention the sorts of questions later raised in both the United States and Germany about biodiversity, global inequality, and planetary health – faded from view as scientists homed in on physical safety as the sole object of concern.

A second episode, exemplifying a pattern similar to the law lag but this time within bioethics, began with the birth of Dolly the sheep on July 5, 1996, but publicly announced by Edinburgh's Roslin Institute in February 1997 after confirming tests were done. Dolly was the first animal to be cloned with genetic material from an adult mammalian cell. Out of 277 tries, hers was the only live birth, suggesting that the technique of somatic cell nuclear transfer was, for that moment at least, far from foolproof. Nonetheless, the question immediately arose whether similar techniques could one day be applied to clone human beings, and policymakers scrambled to find ethical justification for nipping such research in the bud. In the United States, a day after the story appeared in the *New York Times*, President Bill Clinton turned to his National Bioethics Advisory Commission (NBAC) for a rapid response, and on June 9 the commission transmitted its report to the White House (NBAC 1997). Working under tight time constraints, the NBAC hardly had the luxury of deep reflection. Hurlbut discerns in the commission's thinking a pattern of reductionism that occurs again and again not only in encounters between science and law (see Chapter 3), but also in American bioethical deliberation. Science and technology are cast as inherently progressive vehicles of human betterment, whereas social institutions, including bioethics, are seen as lagging, striving to keep up with exciting visions of the future opened up by technoscientific advances. This means that democratic institutions have to take their cue from science to frame the scope and terms of deliberation: "Democracy must delegate to science the authority to define the focus of public debate, and must defer to scientific accounts of what risks and benefits are realistic, before asking whether such risks and benefits are bad or good" (Hurlbut 2017, 145). The language of fact then becomes in effect the language of ethics, and public moral imaginations are reduced to the boundaries of the possible as determined by the sciences of the day. The NBAC, in particular, concluded that cloning to create a child should not yet be undertaken because, on then current scientific information, it was not safe. This assessment avoided immediate controversy while leaving the door open for a very different result if and when science found the procedure to be adequately risk-free.

A third episode illustrates yet another strategy of seeking deference, which is to deny the public its own moral language because it is too ignorant, unscientific, and irrational. This view of an unruly, obstreperous public, driven by emotion not reason, emerged during the Obama years in the deliberations of the Presidential Commission for the Study of Bioethical Issues (PCSBI), chaired by the distinguished political philosopher and university president Amy Gutmann. In its report on synthetic biology, the commission took pains to lay down the preconditions for respectful democratic deliberation, attentive to the diversity of perspectives involved. Key to deliberative success, the commission held, was "information accuracy," a goal to be realized through an independent "fact-check mechanism" that would cleanse the public sphere of dangerous linguistic inaccuracies and distortion. The commission implied that censorship might be needed to keep the public language "clear and accurate":

The use of sensationalist buzzwords and phrases such as "creating life" or "playing God" may initially increase attention to the underlying science and its implications for society, but ultimately such words impede ongoing understanding of both the scientific and ethical issues at the core of public debates on these topics. (PCSBI 2010, 156)

While the idea of a fact-check mechanism to control public language seems almost comically misguided, the PCSBI's recommendation only rendered explicit and concrete ideas about the unruly public that had been circulating through the scientific community for years. One sees the same fear of runaway public discourse in British reproductive scientists' assumption (largely unwarranted as it happens) that science fiction in the form of *Frankenstein* or *Brave New World* had unduly colored people's perception of technologically assisted reproduction (Mulkay 1997, 116–30). A similar anxiety gripped the advocates of California's Proposition 71, a referendum to secure state funding for hESC research. Hurlbut (2017, 214–17) describes the protracted terminological disputes to reach a formulation that would not lead citizens astray or impose an unfortunate "emotional overlay" on wording that should be value-free and factual as opposed to value-laden and political. And hidden away inside the invisible dynamics of institutional research ethics bodies of the kind I myself serve on at Harvard are continual debates about the right way to name things (like SHELEs or SHEEFs or synthetic embryos) so that publics will not get the wrong idea about research, become anxious about non-issues, and, in the worst case, turn off their invaluable support for funding science.

A fourth rationale for deference comes from the narrative of inevitability that often accompanies rapidly moving frontiers in the sciences, as in the case of CRISPR (Clustered Regularly Interspaced Short Palindromic Repeats), or CRISPR-cas9, and similar gene editing techniques in the early decades of the twenty-first century. CRISPRs are DNA segments derived from bacteria, recognizable by the stutter of short, repeating sequences, that can be used to target and modify any DNA sequence that researchers wish to modify, for instance to excise a disease-causing mutation and replace the harmful segment with a healthy allele. The technique's discovery and subsequent rapid development generated huge excitement because it opened up possibilities for manipulating and "correcting" DNA sequences more efficiently, with fewer steps, and at far lower cost than earlier gene modification techniques. Scientists and industry foresaw untold applications in health and agriculture, perhaps to the point where human intervention will indeed be able to redirect the normal processes of evolution (Doudna and Sternberg 2017).

When such revolutionary technological frontiers open, the presumption often is that the science will move forward no matter what law, policy, or democratic debate may do to slow down inquiry. Paradoxically, this belief breeds a sense of urgency together with a commitment to incrementalism. Since science is moving quickly, as recognized for instance by the US National Academies committee on human genome editing (NASEM 2017), the need to take an explicit position seems pressing. And yet the hurried march of science makes the future feel so uncertain that it seems reasonable to let science move on at its own speed and direct one's attention to mitigating risks reactively only as they appear. That default incrementalism, which in effect lets the progress of science and technology determine which issues are ripe for society to evaluate, has marked almost all major national bioethics reports in the United States since the President's Commission for the Study of Ethical Problems in Medicine (1982) issued *Splicing Life*, its report on human genetic engineering. In this respect, bioethics has joined hands with the law in allowing science and technology to seize the moral high ground and to colonize our collective visions of human futures before any other forms of reflection are even allowed to begin.

Conclusion

The biological sciences influence our understanding of life not merely by uncovering hitherto unseen facets of what life is, nor by isolating the material components of living things, nor even by creating new living entities – but equally, and simultaneously, through a control of language and discourse that delimits how we name biological entities and in what terms we debate their meaning or worth. Biology's successes in this respect are partly a matter of relative power. With its call upon the values of objectivity and truth, science commands a repertoire of sense-making that claims to transcend the local and the particular, the subjective and the affective. The color red may mean different things to different people, even when one is not color blind, but an embryo, if accurately defined, should mean the same thing to everyone, regardless of what language one grew up speaking. It simply is what it is. The only permissible destabilizing moves then may come from science itself, as, for instance, when an embryo gets divided into discrete and identifiable entities corresponding to different stages of development – such as after fourteen days, or the formation of the primitive streak, or self-organization into embryo-like form, or implantation in the womb. As the producer-in-chief of novel entities, science asserts the godlike right to name its own creations, without acknowledging that names are not innocent. Under the mantle of mere identification there operates a force strong enough to shape our ability to speak of, and so to govern, the things that form our world.

That naming power, as grand and infinite as Adam's freedom in the Garden of Eden, has seeped, as we have seen, into a growing claim to control the terms of ethical and political deliberation on human futures. Sometimes, as in the debate on embryos and pre-embryos, SHELEs or SHEEFs, science claims unique authority to name, and hence to define and describe, the objects under discussion. Other times, it is a more diffuse power to delimit the framework of debate by excluding matters of concern that do not yet exist in scientists' vocabularies. Outlaw notions that publics or the media may generate (like "creating life") are dismissed as science fiction, or as the emotional overlay on fevered public imaginations, to be reined in through disciplined fact-checking and education. Integral to this set of stylized moves is a sense, shared across other professional domains such as law and bioethics, that normative issues can only be resolved on the basis of accurate factual description. If, and only if, we get the facts of the matter right, there can be a shared reality that would limit, if not dispel, the very grounds for disagreement. Thus, if a growing mass of cells in a dish, for example, cannot ever take on embryo-like characteristics, then the need to limit research on such entities falls away; they are not on a continuum with human life, and hence lie outside the zones of moral concern, until they begin displaying human-ness. But only science, in this way of looking, can tell us if the original ontological boundary was real or imagined, and when if ever a new boundary needs to be set in place.

Science, to be sure, is never alone in crafting society's deliberative discourses and the discursive struggles described in this chapter involved, in all cases, interactions between science and other institutions that also speak with authority, such as legislative committees, bioethics bodies, and on occasion citizens when they choose to challenge science's dominion. The debate on Harvard geneticist David Reich's use of words like race and population (see Chapter 1) is a prominent example. Throughout the history of deliberation on the future of biotechnology, however, we see other institutions repeatedly deferring to science's accounts of foreseeable and desirable futures, and so buying into linear accounts of progress that sideline other possible narratives of hope or fear. Discursive narrowing reduces negative consequences largely to concerns for physical health and safety, well enough governed by discourses of formal risk assessment, so that larger moral questions such as those raised by a Chargaff or a Sinsheimer are banished from regulatory regimes that consider closed doors and lab coats sufficient safeguards for humanity. Positive futures, in parallel, are frequently defined in relation to individual misfortune, with the figure of the genetically doomed patient commanding possibly disproportionate deliberative space in relation to other benefits and beneficiaries. Framings such as the ethics lag, unruly publics, and inevitability further ensure that science gets to define which issues are ripe for debate and in what forums. The power of scientific knowledge seems then, in good part, to draw strength from self-abnegation on the part of other knowledge systems, whose retreat leaves science as the undisputed winner in the language games that attempt to make sense of life. In the next chapter, we turn to an ambitious effort to reorient the discourse of the life sciences to a yet more encompassing vision of global futures – with mixed implications for humanity.

6

A New Biopower

From the earliest days of the genetic revolution, biological research has been justified as a means of preserving and enhancing life, not merely for the therapeutic purpose of curing disease, but also to fight infertility, feed the hungry, diagnose the causes of disability, and at the limit ward off death itself (Horn 2018). Those promises grew more ambitious with the turn of the millennium, a period of stocktaking in which, as we saw in Chapter 4, the imagination of pure discovery yielded to that of engineering life, and biologists began rethinking the methods, directions, and purposes of their field.

Biology, many of its practitioners felt, was due for a face lift, possibly even a disciplinary makeover. In 2004, Carl Woese, a microbiologist at the University of Illinois, wrote a manifesto-style article, boldly titled "A New Biology for a New Century." Here, he decried the reductionist imprint that the turn to physics had left on twentieth-century biology. That move, Woese said, had unmoored biology from its holistic urge to understand life in its fullness: "But the physics and chemistry that entered biology (especially the former) was a Trojan horse, something that would ultimately conquer biology from within and remake it in its own image. Biology would be totally fissioned, and its holistic side would be totally quashed." Thus hollowed out, Woese lamented, biology had ground to a halt as a fundamental science, "its vision of the future spent, leaving us with only a gigantic whirring biotechnology machine." The way forward lay in biology reclaiming "nonlinear" phenomena such as emergence, evolution, and form, "to understand the world, not primarily to change it" (2004, 185).

While others shared Woese's desire to revamp biology's programmatic vision, the notion that this revisioning demanded a retreat from technology was not widely shared. Indeed, the program of synthetic biology, which also gained momentum during the first decade of the new century, embraced the very engineering instinct that Woese repudiated as having no future. Focusing on molecules and cells as building blocks with which to create living things from the ground up, synthetic biology took its inspiration from Lego, the wildly popular Danish children's toy that uses small, standardized components in primary colors to build everything from cars to castles and rocket ships. One key to the toy's success is that the components can be fitted together in an inexhaustible variety of forms, subject to few limits but the designer's imagination. Another is the simple, modular structure of the classic Lego brick that allows any complex form to be broken down – as if in a Cubist image – into an array of interlocking pieces that make combination and recombination feel like child's play. Synthetic biology sought to emulate Lego's success, only with biological components that would add up to new forms of life.

Universities and research funders began to make room for this engineering mindset, which also promised gains for the economy. At Wisconsin, James Thomson's enterprise, the first to derive human embryonic stem cells, was rehoused in a new, largely privately funded home, the Morgridge Institute for Research, which promises "fearless science" aimed at improving human health. At Harvard, a new Department of Stem Cell and Regenerative Biology was formed in 2007, incorporating faculty from multiple schools across the university; a companion Harvard Stem Cell Institute leveraged private funds to translate research from lab to drug discovery. Also in 2007, MIT, the leading East Coast center of engineering innovation, formed a Department of Biological Engineering,[1] staffing it with young researchers, some of whom openly broke with the project of molecular biology.

One of those renegades, Drew Endy, who later left MIT for Stanford, spoke enthusiastically of the expanding prospects for "engineering biology" in a 2008 interview with *Edge*, the online platform for the self-styled "third culture." Older biotechnology, Endy claimed, had fallen short in all three tasks it had embraced in its early years – to make new therapeutic drugs, initiate gene therapy, and engineer plants that do not need synthetic fertilizers – and in three decades only the first goal was accomplished. It was time, Endy insisted, for a strategic shift: from the messy mapping, isolating, and rebuilding methods of molecular biology to something quicker, simpler, and more accessible. He concluded: "Thirty years into biotechnology, despite all of the successes and attention and hype, we still are inept when it comes to engineering the living world. We haven't scratched the

144

surface of it, and so the big question for me is, how do we make biology easy to engineer?"[2]

This was heady language. The US National Science Foundation (NSF), mindful of the next big thing that might justify substantial public expenditures, seized on this vision to ramp up a new initiative in synthetic biology (Roosth 2017). A 2006 NSF grant established Synberc, a multi-university research center built on Endy's mission statement, "to make biology easier to engineer."[3] Despite its commitment to building "foundational tools and technologies," the center's primary aim was to produce solutions, especially in the areas of heath care, energy, and environment. In other words, biologists who embraced the engineering turn in their field saw themselves – as engineers have since the beginnings of human history – as problem-solvers, pursuing knowledge not for its own sake but to serve larger social purposes.

Gene editing, with its promise of transforming diagnostic, therapeutic, and agricultural practices through targeted interventions, beckoned toward additional realms of possibility. Eradication of stubborn and harmful biological species, such as the malarial mosquito, suddenly seemed feasible. Using CRISPR, one could create less fit versions of the offending organism and then use another naturally occurring biological system, gene drives, to propagate the edited version quickly throughout a population. Eventually, in theory, such interventions could cause a harmful population to crash, as enough members were replaced with sterile variants or with variants that would produce only one sex of offspring in the next generation. At MIT, Kevin Esvelt, a pioneer in combining gene editing and gene

drive technologies, seemed to speak for a new generation in merging technique with morality. Scientists, he
said, become morally responsible for ending suffering
when they acquire the power to intervene.[4] Such a claim
presupposes that, along with the power to change the
structures of life, comes the wisdom to discern which
problems most need correction and whether it is right to
alter nature selectively to achieve those ends.

Apostles of bioengineering like Drew Endy and Kevin
Esvelt did not see the ambition to change the world as
a barrier to understanding, as Woese did. If anything,
engineering simply provided for them a more hands-on
path to improving life, by modeling and mimicking it,
by correcting its errors, and by altering its courses when
science finds naturalness too restrictive or, as Esvelt
insists, too cruel. Bioengineering's focus on solutions,
however, elided an important threshold question, possibly the most important question in a democracy. Who
defines the problems and, in cases of competing goals,
who sets the priorities? For biology in the twenty-first
century, the answer, increasingly, appeared to be not the
people, but scientists and engineers, because they best
understand what science and technology can achieve.
And legal, political, and bioethical institutions, as we
saw in previous chapters, generally deferred to these
agenda-setting moves on the ground that discovery and
its applications are synonymous with the public good.

A 2009 report from the US National Research
Council (NRC), the policy analytic arm of the National
Academies of Sciences, Engineering, and Medicine, can
be seen as an authoritative exposition of this imperial
vision, one integrating an expansionist agenda for the
biosciences and technologies with broad emancipatory

promises for society (NRC 2009). The report is worth dissecting in detail for that reason alone, but that is not all. This was not a random review of the state of the biosciences. The report was commissioned by three federal funding agencies – the NIH, the NSF, and the Department of Energy – to recommend how the nation might "capitalize on recent technological and scientific advances that have allowed biologists to integrate biological research findings, collect and interpret vastly increased amounts of data, and predict the behavior of complex biological systems." It informed the 2012 US National Bioeconomy Blueprint, a call to harness the life sciences to promote economic growth.[5] It is in these respects not merely an isolated report, but an important expression of a larger imaginary of science-based governance – one in which US leadership in science and technology translates none too subtly into ruling prescriptions for how to manage the world in its entirety.

A revolutionary mission

Titled very similarly to Woese's article,[6] but patriotically subtitled *Ensuring the United States Leads the Coming Biology Revolution*, the report – *A New Biology for the 21st Century* – laid out in five chapters the ways in which biology is poised to deal with some of humanity's grandest challenges, especially in the realms of food, environment, energy, and health. This program, admirable enough in its objectives, neatly illustrates how science's monopoly on declaring what life *is* continually segues into judgments about what life is *for*, in short, into configuring the directions of human progress.

147

With its self-conscious use of capital letters, the NRC report signified the New Biology's break with its past – from a record of illustrious discovery to a future of governance and control. Biology's new research agenda, as the authors conceived it, has two salient characteristics that justify its ambitions and its demand for increased public resources. First, it is a product of cross-disciplinary convergence, and hence it is biology plus, even biology on steroids: "the essence of the New Biology is integration – re-integration of the many sub-disciplines of biology, and the integration into biology of physicists, chemists, computer scientists, engineers, and mathematicians to create a research community with the capacity to tackle a broad range of scientific and societal problems" (NRC 2009, 3). Second, this hybridized field, centered in the life sciences, has a special claim on public funds in the committee's view because it is "purpose-fully organized around problem-solving"; it represents in this respect an "additional, complementary approach to biological research." Overall, the "committee recom-mends setting big goals and then letting the problem drive the science" (2009, 5). The New Biology approach thus offers a paradigmatic example of the kind of out-ward-directed science that European analysts identified in the 1990s as Mode 2 (Gibbons et al. 1994). It is interdisciplinary, interinstitutional, context-dependent, and unabashedly action-oriented, a far cry from the detached, curiosity-driven, discovery-for-its-own-sake science of the cloistered ivory tower. But whereas in the classic Mode 2 model agenda-setting is imagined as a public process, here, through a subtle elision of the pure–applied boundary, biology takes on that guiding role for itself.

A New Biopower

While promising a wealth of technological solutions, the New Biology still trades on the authority of pure science to retain control over its research directions and priorities. Problem-focused though it is, the NRC report stresses the gains in basic understanding that are expected to flow from the proposed initiative. For the report's authors, as for Carl Woese, making sense of the complexity of life is presented as the most important goal, and the sorts of problems they have in mind encompass life in its widest manifestations. Under the heading of food, the New Biology aims to develop knowledge of efficient plant–environment interactions "under different conditions." Environmental research would unite engineering and computer science with biology to produce better understandings of ecosystem services and biodiversity. The energy focus would advance "fundamental knowledge" as well as tools and technologies to produce plant-based alternatives to fossil fuels. And health research would unravel the "interacting networks of staggering complexity" that lie between "the starting point of an individual's genome sequence and the endpoint of that individual's health" (NRC 2009, 4–5).

To paraphrase Dorothy in *The Wizard of Oz*, we are not in Schrödinger's world anymore, that world where the regularity as well as the diversity of life could be traced back to a central control station and the application of a single master code-script. There are, to be sure, surface similarities between the physicist's project of decoding life and the New Biologists' project of unraveling complexity. Both wish to find the primary levers that control pattern and regularity in living systems. But the differences are far more striking.

149

Against Schrödinger's focus on reducing the structure, functioning, and reproduction of living organisms to the molecular level, and reading everything in terms of a single script, the New Biology seeks to understand life within, and in interaction with, ever-changing contexts and environments. Whereas the eye of the physicist burrowed deep into the insides of cells and chromosomes to pinpoint the sources of regularity in genes and cells, the New Biology draws the entire world into its investigative ambit. And this New Biology's avowed purpose is regulatory – engineering solutions to problems of living that originate in society rather than finding answers to puzzles embraced by science for its own enlightenment. This is biopower on a scale that Michel Foucault, who coined that term to describe modern states' "calculated management of life" (1990 [1976], 140), would have found astonishing. It is biology on the path to planetary control.

Bigness entails its own simplifications. Many things are taken as given in the account of society and its problems that the New Biology of the NRC report sets out to ameliorate. There is, to begin, a presumption of universalism in the framing of issues: that the problems of food, energy, environment, and health are the same for all humankind, and therefore solutions too will follow from a single revolutionary research agenda. There is thus no perceived contradiction between maintaining global leadership in the development of biology within the United States and solving worldwide problems: New Biology for one is New Biology for all. Further, the report presumes that the sciences and technologies whose integration makes up the New Biology can make progress on their own, without involving the social and human sciences, not to mention law and humanistic thought. The implicit assumption is that big problems may originate in the social world – such as overconsumption or scarcity – but solutions will be found in altogether different fields of endeavor, with no need for exchange between the producers of New Biological understandings and the eventual consumers of its proposed solutions. Third, the vision of progress outlined in the New Biology report makes little or no concession to possible value conflicts among interest groups, across nations, or among divergent philosophical schools about the endpoints of biological intervention or concepts such as "wise stewardship of our planet" (NRC 2009, 2). In ethics as in science, the world is taken as flat, with everyone presumed to value the same outcomes to the same degree. Last but not least, scientific advances are thought to include their own checks and balances, with no need for public accountability to monitor the pace, the directions, or the uses of new biological knowledge.

These are not surprising omissions from a report written by scientists for their own nation's science policymakers, who also hold the purse strings for research funding, but what looks uncontroversial from the viewpoint of elite (and in this case also nationalist) science is not equally so from other perspectives. Under the aegis of solving problems, the authors of the report in fact lay out a particular charter of human progress that appears so self-evident to them that it cannot be questioned any more than one can question the laws of nature. Yet, issues not seen as significant, indeed not seen at all, in the 2009 NRC report were percolating over decades through the deliberations of ethics commissions and regulatory bodies, and surfacing in both academic writing and public controversies. In the following sections, we look more closely at the disconnects between the New Biology's aspirations for global problem-solving and the issues of governance highlighted by scholars and critics of biotechnology both within and beyond the United States. A brief comparison with a roughly contemporaneous UK report on emerging biotechnologies rounds out our understanding of the ways in which the NRC's discourse on public benefits narrowed the questions for deliberation while forecasting a glowing future for American biology and its worldwide publics.

Feeding the world

Among the grand technoscientific projects of the twentieth century, few loom as more politically loaded than the effort to feed an increasing global population that seems at times to be hurtling toward Malthusian disaster. By mid-century, population growth in the world's poorer regions seemed to be outstripping agricultural production. Grand geopolitical battle lines were drawn between the Red tide of socialism and a nascent Green Revolution catering to the needs of developing countries – warding off starvation, mass social unrest, and political upheaval. With assistance from the Rockefeller Foundation, Norman Borlaug, an American agronomist born in the Midwestern farmlands of Iowa, embarked on a project of improving plant life to increase the yield of staple crops, in particular, wheat in Mexico.

According to the near-canonical story, the high-yielding wheat variety that Borlaug helped develop first took root in Mexico and then in South Asia and beyond, where similar breeding techniques were also applied with great success to rice. Disastrous food shortages, such as the Great Bengal Famine of 1943 (Sen 1981) that took more than two million lives, receded into history as newly decolonized nations like India moved from cash-strapped dependency on imports to becoming net exporters of grain. Science, specifically agricultural science, according to this narrative, turned the page from poverty to plenty, proving its capacity to liberate the world from hunger. Borlaug himself is sometimes credited with saving more lives than anyone else in human history, although through a long working life he wore his celebrity lightly.

Borlaug's project was rooted in another period of scientific invention, from the 1920s to the 1950s, that the historian Lily Kay (1993) also referred to as the rise of a "new biology." Then, too, biology turned self-consciously interdisciplinary. Spurred by the ministrations of the Rockefeller Foundation and Caltech, methods were borrowed from physics, chemistry, mathematics, and other life sciences to create an integrated "molecular vision of life."[7] There were fundamental differences, however, between the New Biology of the twenty-first century and its less highly capitalized predecessor, the new biology of some decades back. That earlier, simpler synthesis was still bounded by the contours of the organism to be modified. It "narrowed its principal focus to macromolecules," based on the "premise that the salient features of life – reproduction, growth, neural function, immunity – could be explained through the structures and functions of proteins" (Kay 1993, 5). In time, this "protein paradigm" would help create a "new science of

153

life, a science whose tributaries eventually converged on the molecular study of the gene" (1993, 6).

Understanding genetic variance guided Borlaug and his research team to breeding the squat, sturdy plants that account for the Green Revolution's impressive gains in yield. The breeder's trick was to shrink plant height using so-called dwarfing genes, while still supporting the increased grain weight produced through intensive irrigation and fertilization. This was Mendelian genetics turned to life-saving use: from the standpoint of food production, the results were little short of miraculous. But critics soon pointed out that those gains had come at environmental and social costs that were never factored into assessments of the Green Revolution's benefits. Political impacts, beyond concerns about Cold War polarization, such as interregional economic disparities and social inequality, likewise remained out of sight. In time, critics from the global South, personified for many by the charismatic Indian activist Vandana Shiva (1991), blamed the Green Revolution's limited vision of plant improvement for longer-term destruction of soil quality, threats to biodiversity and food security, and takeover of farmer autonomy through corporate control of seeds. Shiva herself was often reviled for emotionalism and unscientific excess (Specter 2014), but even promoters of the original dwarf wheat and rice varieties conceded that a more systemic approach to environmental impacts might have alleviated problems that the revolution's scientific leaders overlooked (Gillis 2009).

Third World environmentalists and social critics were not the only ones resisting agricultural biotechnology. Britain in the 1990s initiated a long-drawn European struggle against American genetically modified (GM) crops. The resulting conflict among economically and technically equal adversaries brought to light important scientific uncertainties and social ambivalence that were prematurely sidelined in the US rush to adopt a new, commercially beneficial technology (Jasanoff 2005). A trade dispute at the World Trade Organization appeared to give the United States and other GM grower nations a formal victory, but failed to convince European nations to go along with a particular hegemonic vision of industrial agriculture that did not sit well with local political and social values (Winickoff et al. 2005; Jasanoff 2006). Meantime, with global populations still rising, leading biotech companies, most notably Monsanto, and their academic advocates railed against Europeans for imposing their rich-country preferences on a world that could ill afford the luxury of doing without advances in scientific agriculture (Paarlberg 2008). Many claimed this position was not merely unscientific but downright immoral (Daniels 2017).

That assessment is premised on a governance model for biotechnology that can be termed the Asilomar consensus: namely, if physical safety concerns are adequately addressed, then regulation can reasonably be left to expert judgment. As a corollary, if people reject scientific conclusions about what industrial biotechnology can or should be allowed to do, it must be because they do not understand the science and should be re-educated, as Maxine Singer argued (see Chapter 3). In many parts of the world, however, this analysis came across as intellectually and ethically wide of the mark, even bankrupt. Although Asilomar's success lodged itself in scientists' memory as an almost indisputable fact (Hurlbut 2015), the world in effect spoke back in vigorous denial. The 1975 consensus was too narrow, too specialized, possibly premature, too insensitive to local conditions and sensibilities, and too oblivious to moral concerns about managing nature.

What is at stake in these debates is not knowledge of science but acceptance of its universalizing ambitions. If Borlaug thought a revolution in plant breeding could feed the whole world, then Monsanto's claim that a genetically standardized, perfected wheat would grow anywhere, and therefore should be grown everywhere, carries that argument to yet less defensible extremes. It overlooks the environmental, economic, and social detriments that accompanied the Green Revolution, and it entirely sidesteps the political questions associated with a revolution in food production spearheaded not by scientific altruism, such as Borlaug's, but by monopolizing capital. One can see in the decades of global resistance to GM crops a demand for new forms and forums of political debate on biotechnology's promises of a "doubly Green Revolution" (Conway 1999). The New Biologists' hope of regaining the high ground through a renewed science-propelled drive to feed the world seems not to have ignored those lessons completely.

Strikingly, the New Biology report's depiction of new horizons in crop plant research makes no mention of economic and social infrastructures or political conflicts. The focus is on ecosystems in which plants interact with diverse physical and biological parameters, ranging from rain and sunlight to insects and birds, but evidently not humans. There is no mention of the implications of farm size for agricultural improvement, or the effects of regional disparities in income or in farming cultures, no indication that biotechnology advances only in concert with private capital and intellectual property rights, and no admission that biological research decoupled from the human and social sciences will never be able to meet society's demands for accountability. As we will see below, each of the other ambitious improvement projects laid out in the report displays a similar blindness to politics and values.

A measured environment

The New Biology's environmental promises home in on preserving ecosystem function and biodiversity. Here the hope for sustainable life on the planet rests on the power of quantification to measure how well the environment is serving humanity's needs and delivering what the report, borrowing from environmental economics, calls "ecosystem services" (Costanza et al. 1997). The 2005 Millennium Ecosystem Assessment (MA) defined this term simply as "the benefits people obtain from ecosystems" (2005, v). These are further broken down into *provisioning, regulating, cultural,* and *supporting* services, referring respectively to providing sustenance, affecting climate and other dynamic systems, offering recreational and esthetic opportunities, and underpinning basic processes of life, such as photosynthesis. The New Biology report takes the MA a step further, promising to integrate knowledge across vast frontiers "through the unifying languages of mathematics, modeling, and computational sciences" (NRC 2009, 25). Aided by these technical instruments, biology promises to develop indicators of degradation, monitor ecosystems, identify risks, and even restore lost function.

The idea of a science that can predict and ward off catastrophic events is deeply reassuring at a time when the Earth is experiencing many macroscale threats – a trillion-ton iceberg breaking off the Antarctic ice shelf, Bolivia's Lake Poopó drying up, megacities like Jakarta sinking into the sea. Yet, a report card on management based on the notion of ecosystem services would look far from spotless. Justification for particular policy approaches often builds on examples that have acquired the status of truth through repetition and amplification, and yet seem shaky on closer inspection. One frequently cited case is that of New York State's restoration of the Catskills watershed at a cost of about $1–1.5 billion. That investment allegedly allowed "natural capital" to perform needed storage and filtration services, saving the city up to $6–8 billion over ten years in building, operating, and maintaining artificial water treatment facilities (Chichilnisky and Heal 1998). But is this parable based on historical evidence or on wishful thinking? Mark Sagoff, philosopher and frequent critic of economic approaches to valuing the environment, strongly argued that wishfulness prevailed. He observed that New York's water quality was neither seriously threatened nor in need of more treatment; and despite an agreement with the US Environmental Protection Agency, the city did not actually invest anything like the amount of money in land acquisition or nature preservation that the parable claims. If the story continues to persuade policymakers, it is because:

> Few of us wish to admit that we benefit from nature not by preserving but by "improving" it – for example, by plowing a field, building a road, constructing a house,

drilling a well, damming a river, farming a salmon or oyster, or altering a genome. Most of us would rather believe that Nature knows best. We will therefore repeat ... the Catskills parable even though it is factually false. (Sagoff 2005)

To turn ecosystem services into an instrument of environmental governance, moreover, requires the creation of markets, and this step too is fraught with difficulties. A study of three such markets, in the Chesapeake Bay watershed, the Ohio River basin, and the Willamette River basin, concluded that part of the problem lies in the very attempt to disaggregate and quantify elements that are closely bound up together – for example, how much water or fish a river basin supplies and what such a place means to people esthetically or for a sense of community (Maasakkers 2016). Further, even if indicators could be devised to accurately measure the disparate services provided by ecosystems, the very act of turning the varied meanings that people attach to nature into a single set of numbers can prove to be misguided. Wendy Espeland (1998) documented how a Bureau of Reclamation project to build a dam in Central Arizona brought to light divergent, and incommensurable, values attached to the affected land by the bureau's efficiency-seeking engineers and economists and the Yavapai Indians who would have lost their ancestral lands to flooding. Such differences, Espeland and other sociologists of knowledge have argued, cannot be merged into a single calculus of rationality. To force such mergers, as in the NRC report's advocacy for greater integration and coherence across indicators for assessing ecosystem performance, is to ride roughshod over human values.

Energy imaginaries

The New Biology's solutions for the world's energy challenges center on the need to find sustainable alternatives for fossil fuels, and identifies biofuels, such as ethanol derived from grains, as the primary candidate. In flagging this as a grand challenge, the leaders of science essentially echoed governmental policies of the period. Soon after the turn of the century, both the United States and the European Union enacted rules to increase the fraction of fuels from plant-based sources in the transport sector in an effort to cut greenhouse gas (GHG) emissions. Crop plants were massively pulled into service to meet the transportation needs of increasingly mobile industrial societies. In the United States, cornstarch provides the primary raw material for ethanol production, consuming about 40 percent of the corn grown in the country. This amount was thought to have plateaued by 2016 because human food and animal feed account for the remainder of the crop (Hood 2016). Accordingly, it made sense for biological research to seek alternative plant-based sources and to aim for more efficient conversion of biomass to biofuels.

Once again, however, the New Biology's engineering inclinations left human behavior and human needs out of its research program, ignoring problems arising in human–nature interactions. These surfaced quickly. A decade or so into the new era of mandated biofuel production, it became clear that an optimistic, productionist framing of the problem had overlooked how behavior might change in response to state biofuels directives. Chief among these was the phenomenon of

160

"indirect land use change," transforming both cropland and non-cropland, such as forests and grasslands, for biofuel production. Changes in vegetation patterns and associated agricultural practices increased overall GHG emissions enough to more than offset the reductions achieved from switching to biofuels. As in the case of the Green Revolution, governmental decrees stipulating what kinds of plants should be grown unleashed a variety of secondary, but altogether predictable, consequences that were not foreseen in the original hopeful calculus of swapping out fossil fuels with biofuels.

Another imperfectly foreseen impact of biofuels production was an increase in global food prices, as vegetable-based oils were diverted for transport use. Yet, this result was entirely in keeping with fundamental economic rules of supply and demand. Indeed, a 2017 review concluded: "The overwhelming consensus in the literature we surveyed is that, as predicted by basic economics, biofuel demand (and hence biofuel policy) results in increased food prices" (Malins 2017, 3). Policymakers were forced to take note: in 2015 the European Union amended its 2009 Renewable Energy Directive to cap the share of food-based biofuels to 7 percent of total supply,[8] and recognition grew that this cap, too, might have to be substantially lowered. The surprise, if any, is not in these sorts of linkages between policy and its effects on consumption. Rather, it is in the policymakers' capacity to overlook the predictions of "basic economics" in the zeal to pursue technological fixes. And it is in the correlative presumption that biologists alone – even under the broad systems perspective of New Biology – can solve the world's energy problems by focusing only on plants and their components, while the

161

people, cultures, and habits that make up the economic and social contexts of food production and land use remain out of the analyst's purview.

The networked genome

The report's fourth focus on health as a key "societal challenge" was a foregone conclusion. From its earliest beginnings, the biotechnology industry had dangled the prospect of cures for untreatable diseases before an expectant society, and the search for treatments acquired new momentum with the completion of the Human Genome Project. The map of the human genome, coupled with increased computing power, opened up vast new possibilities for identifying patterned genetic variations connected with disease. Instead of looking for needles of useful data within haystacks of undecipherable information, a more targeted science of bioinformatics seemed poised to find meaningful signals corresponding to manifested disease conditions. Governments and private businesses began constructing biobanks, storehouses of genetic data from large populations, in an effort to locate the complex linkages involving multiple genes that may account for hard-to-treat conditions such as mental illness and degenerative diseases of the brain and nervous system. In 2014, Steven Hyman, a former director of the National Institute for Mental Health, optimistically declared that genetic insights had "transformed a featureless landscape into one with real scientific toeholds," and concluded, "I am convinced that genetic variants for depression can be found."

In framing the health research tasks for New Biology,

the report zeroed in on a parallel between ecosystem monitoring and restoration and "individually predictive surveillance and care" for the human body (NRC 2009, 31). This is the goal of precision medicine, a vision of health care that imagines each human being as a uniquely delineated map of genetic predispositions and phenotypic (i.e., directly observable and measurable) factors. As if the genotype–phenotype nexus were not complicated enough, biologists also homed in on the microbial communities, or microbiomes, that live within the human organism and regulate its performance in unpredictable ways, for example producing obesity or leanness in identical twins endowed with the same genetic characteristics. Genomics, in this respect, opened up an expanding frontier for biological investigations connected with health. At the same time, it also invited the biologists' gaze yet deeper into the inner workings of individual human bodies and their networked genomes.

Many of the largest health gains of the past few centuries, however, have come from better understanding of the environmental determinants of health and increased knowledge of how humans behave as social animals. Improved nutrition and sanitation in the nineteenth century led to spectacular gains in life expectancy, as diseases like malaria and cholera disappeared from much of the prosperous industrial world. Even before vaccines, strict quarantining practices reduced the threat of many childhood infectious diseases, from measles to polio. Closer to our time, anti-smoking campaigns and regulations from the 1980s onward coincided with significant drops in lung cancer rates and heart disease. Around the world, human life expectancy increased markedly over the past century, but these gains had at least as much to do with cleaner environments, healthier eating habits, and increased physical activity as with pharmaceutical interventions, even the widely used cholesterol-reducing statins. Put differently, public health measures brought gains to human lives on a scale that genetically informed precision medicine can, as yet, only approach in biologists' dreams.

Treating the genomic human as a "web of networks" seems at first a far cry from the close-focused, reductionist study of genetic variation in the early days of the new biology described by Kay and others. Clearly, biologists now recognize that common human illnesses such as diabetes or heart disease involve many more systems and parameters than defects in single or even multiple genes. Yet, delving into genomic complexity risks bringing its own reductions by leaving out of health research the interactions between human beings and their physical and social environments that have produced such large swings in population-wide health outcomes. Smoking-induced cancer, for instance, rose and fell not mainly through the accrual of genetic knowledge or therapies, but because people decided en masse to quit smoking. On the minus side, the collapse of public health systems, the breakdown of environmental standards, the rise of social isolation, and habits such as addiction can worsen basic health outcomes even in advanced industrial societies (CDC 2018). By overlooking these determinants of health, the New Biology approach may replicate the big erasures and blind spots that have compromised the life-enhancing promises of modern biomedicine.

A New Biopower

A more open future

How, then, should we live with the promises of biotechnology on a bounded planet with changing human needs, expectations, and desires? That question has perplexed science, government, and society from the beginnings of biotechnology, but approaches to answering it have diverged across deliberative cultures, revealing different understandings of how to imagine, and manage, the relations between *zoe* and *bios*, bare life and life in society. Britain's Nuffield Council on Bioethics, for example, issued a report in 2012 entitled *Emerging Biotechnologies: Technology, Choice and the Public Good*.[9] In keeping with British practices of trying to incorporate all relevant viewpoints, the twelve-member committee authoring the report included not just scientists from universities and industry – as did the New Biology committee – but also representatives of bioethics, history, political science, and science and technology studies. Its charge too was broader: to recommend ways forward in research, policy, governance, and public engagement, given the complex social context for biotechnology's development.

Beginning with the report's title, which contemplates a plurality of technologies rather than a single, monolithic science, the Nuffield report stressed the need to include diverse perspectives and allow for the contingencies that may affect social outcomes. Its conclusions reflect a corresponding reluctance to embrace specific futures or trajectories of development. Instead, the report recommends a "public ethics" that explicitly considers the existence of multiple, possibly clashing conceptions of the good and seeks to ensure that premature framings by dominant interests do not rule out alternatives that might increase welfare (Nuffield Council 2012, 61–8). In contrast to the confident, solutions-oriented tone of the New Biology report, the Nuffield document projects a hesitant agnosticism that is professedly pro-technology but against choices that may constrain life's possibilities. It rejects the notion that there could be "a single, rational answer to the question 'what is the good life'?" (2012, 62). Rather, guided by the public values of equity, solidarity, and sustainability, the report recommends a cautious openness to more than one arrow of progress:

> We observed that the possible pathways for a range of options cannot always be seen clearly in advance, especially where certain preferred technological pathways are assumed to be urgent and without alternatives. We therefore recommended a more circumspect approach in which commitments to particular technological pathways should be evaluated not only in terms of their expected future impacts but also by comparison to possible alternative pathways. (2012, 175)

Coming from a context of British empiricism and skepticism toward grand narratives, we see here, some thrity-five years further on, a curious convergence with the German Greens' demand in the late 1980s for alternatives to singular visions of sociotechnical progress. We see, too, and still more importantly, a willingness to entertain uncertainty on the issue of the public good – what life is for – that the New Biology report did not reckon as a possibility.

Conclusion

A decade into a new millennium, American biologists embraced the idea of a New Biology as a preeminent scientific instrument for understanding and ameliorating life on the planet. But, unlike the microbiologist Carl Woese, the engineering side of the discipline saw nothing contradictory between exploring complexity and changing the world. Dividing the cares of human societies under four major headings – food, environment, energy, and health – the advocates of an encompassing, integrative New Biology set forth an ambitious research agenda for investigating the complexity of the Earth's living systems and remedying their defects while boosting the economy. Their vision, at one level, was extraordinarily inclusive. The New Biologist did not even have to be a biologist in a traditional sense. Instead, the field spread out a welcome mat for anyone wishing to investigate or intervene in life's myriad aspects: "the New Biology includes any scientist, mathematician, or engineer striving to apply his or her expertise to the understanding and application of living systems" (NRC 2009, 20). Yet, that enticing vision left little space for contributions from outside the natural sciences; even the social sciences were dismissed as an "exciting interface" for future collaboration that this committee would not address (2009, 10).

Omissions, however, have consequences, as is clear from a comparison of the NRC report's confident predictions with the Nuffield Council's more tentative and exploratory conclusions. While conceding that science and technology alone cannot hope to solve all of our food, energy, environmental, and health problems, the New Biology research agenda frames the forward path in ways that ignore humanity's ambiguous experiences with interventions driven by the life sciences of earlier eras. In the triumphalist litany of what science can achieve, elementary contradictions went unnoticed. The report noted neither the potential loss of ecosystem services from converting more of the Earth's arable surfaces to biofuels production nor the effects of increased biofuels production on global food prices. Systems thinking permeates the report's language and proposed solutions. Nonetheless, the systems that the New Biology addresses remain fixated on measurable, physical parameters, failing to take on board how human societies' attempts to understand and make sense of their own condition loop back and reshape the very challenges that confront all life on the planet.

This major restatement of biology's governing vision, articulated from the pinnacle of the American scientific community at the beginning of a new century, performs many of the same reductions and erasures we observed in episodes discussed in earlier chapters. Science in effect claims a monopoly over the steering of human progress, a progress secured through the kinds of engineered solutions that a biology armed with the awesome power of a new bioinformatics can imagine for itself. It arrogates to itself the right to determine what life is for, along with the capacity to discover and redesign what life is. We turn in the last chapter to possible ways of relaxing this monopoly and restoring a more profoundly humanistic sensibility to charting both the ends and the means of progress.

7

Life's Purposes

The renowned Argentinian short-story writer Jorge Luis Borges composed one of his most enigmatic and generative works, *The Library of Babel*, in 1941. The plotless narrative unfolds in a fictional library whose physical organization Borges describes in great detail. It is infinite in reach, yet highly structured, consisting of an endless array of hexagonal rooms, each identically equipped with exactly twenty shelves on four walls, holding "thirty-five books of uniform format; each book is of four hundred and ten pages; each page, of forty lines, each line, of some eighty letters which are black in color" (Borges 1962). A metaphor for the limitless universe of language, Borges's library continues to resonate with readers in our digital age. Among many other influences, it inspired the programmer and writer Jonathan Basile to design a website where the diligent seeker after knowledge can experience the feeling of getting lost in a resource which, according to its creator, now "contains anything we ever have written or ever will write" (Basile 2015).

Borges's story preceded by a dozen years the decoding of the structure of DNA, and yet his description of the library and of the actors scurrying around inside bears uncanny resemblances to biologists' search for the control of life in the genomic age. There is, first of all, the library's sheer immensity; its halls are endless, rendered even more so with reflective mirrors. Just as the human genome encodes infinitely many future human lives, so Borges's library allows for the generation of all possible works of language. Yet, its physical capaciousness is created, as in the human genome, through a series of infinitely interlocking structures built out of a parsimonious vocabulary of shapes and a limited repertoire of letters. Sometimes, the author tells us, those letters add up to volumes with no discernible sense or meaning: "One which my father saw in a hexagon on circuit fifteen ninety-four was made up of the letters MCV, perversely repeated from the first line to the last." These repetitive formations echo the contemporary genomic language with its occurrences of nucleotides arranged in "tandem repeats" and, more recently, the "clustered regularly interspaced short palindromic repeats" that are the carriers of CRISPR science. Then there is the architecture of the whole, held together by "a spiral stairway, which sinks abysmally and soars upwards to remote distances." That spiraling structure, branching into the rooms of the library, seems as if inspired in a premonitory dream by the as yet undiscovered double helix of DNA's sugar-phosphate backbone, holding the eternally paired and bonded bases, adenine to thymine and cytosine to guanine. Indeed, the philosopher Daniel Dennett (1995) capitalized on this resemblance when he likened Borges's Library of Babel to what he called the Library of Mendel, which consists of all possible genome sequences that can be "written" with DNA's basic alphabet of A, T, C, and G.

The force of the analogy, however, may reside less in the structure of the two libraries – the fictional one of literature and the physical model built years later in the Cavendish Laboratory – than in the search for truth and meaning that Borges imagines for those set loose inside his infinitely extended repository. There is the initial excitement of unknown languages waiting to be discovered, as in the "junk DNA" of the human genome whose exact functions still lie outside scientific understanding, much as the "four hundred and ten pages of inalterable MCV's cannot correspond to any language." There is nevertheless a sense of promise and completeness, that "aha" reaction that has marked biology's eureka moments from the initial discovery of Watson and Crick to engineering insights such as Kary Mullis's vision on his moonlit ride down the California coast. Borges says of his library that when it was declared to contain all books, "the first impression was one of extravagant happiness": people felt "there was no personal or world problem whose eloquent solution did not exist in some hexagon." There were even Vindications, "books of apology and prophecy," which justified every person's existence with explanations of past acts and foretold futures. Those forays of enthusiastic discovery do not always end well. Searching for the prophetic books drove some users mad and others desperate, and yet the sheer urgency of the quest seems a foreshadowing of today's race toward precision medicine, tailored to correcting each person's inherited genomic deficiencies and weaknesses. Even the sense of perverted and suicidal ends comports with the darker visions of a few biologists who, like Erwin Chargaff, spoke in a prophetic register of the ends of molecular biology. "If medicine persists in the path it seems to have chosen at present," Chargaff (1987, 200) wrote, "we shall soon hear that its great mission actually is the abolition of death."

That soon is now. The formation of a new biotechnology company named Calico to study aging and lifespan heralded, it was said, Silicon Valley's "obsession with mortality" and its belief "that the day when technology makes it possible to live forever is just around the corner" (Miller and Pollack 2013). More recently, SENS (Strategies for Engineered Negligible Senescence), another Silicon Valley enterprise, has begun an integrated search for fixes against multiple physical correlates of aging that might allow people to live a thousand years – a dream of near-immortality that male tech billionaires such as Palantir's Peter Thiel and Google's Sergey Brin are said to cherish (Horn 2018). Not to be outdone, the National Academy of Medicine (NAM) has announced a Grand Challenge in Healthy Longevity. As recognition dawns that aging may be too complex for singular solutions, the code metaphor remains for some the key to success, if only for fundraising. In the *New Yorker*, Tad Friend (2018) reports Dr. Joon Yun's rallying cry at the Los Angeles launch of the NAM initiative: "'I have the idea that aging is plastic, that it's encoded,' he said. 'If something is encoded, you can crack the code.' To growing applause, he went on, 'If you can crack the code, you can hack the code!'" Horace Judson's fear of the "plasticity of man" has happily given way, it would appear, to a celebration of that very plasticity.

But whose lives will be extended through aging research and what for? Sources such as Calico's website do not reveal, any more than the books in Borges's library, who is destined to live forever as a result of that company's research and development, let alone what kinds of lives they are entitled to expect.

Credo

This book began with a question: can science make sense of life? When I told a colleague this was what I was working on, he replied didn't I know it was a bad idea to ask a question with an obvious answer (I don't know whether his own was *yes* or *no*). But the point surely is that there can be no single, simple answer to that question, because the thing called "life" is capable of infinitely many meanings. No single discipline, even one with the presumed inventive power of the New Biology, can make sense of that word without first reducing life's complexity to a degree that deprives life of the beauty and meaning that make it miraculous.

The philosopher's binary distinction between natural and human kinds gets us part but not all of the way to understanding science's simplifying moves, and seeing what is the matter with them. The philosopher Ian Hacking has argued in many of his works that human kinds, unlike natural ones, are products of our labor in classifying and describing ourselves, or, as he crisply put it, "how names interact with the named" (2006, 23). In the coda to her book *Making Sense of Life*, Evelyn Fox Keller argues that life itself is a human kind. She calls attention to the diverse ways in which biologists have tried to understand what life is through the ages – using models, metaphors, and machines – and each move has yielded a different sense of how to characterize *bios*. The question "What is life?" she concludes, is therefore a historical one, "answerable only in terms of the categories by which we as human actors choose to abide" (2002, 294). But her vision, that of a historian of scientific ideas, is narrowed to only one class of human actors, scientists, who, despite their best efforts, will never fully define, even if they importantly help steer, the totality of human conceptions of life and its purposes.

We have seen throughout this book attempts by the newly empowered life sciences of the twentieth and twenty-first centuries to redefine what is worth understanding about life to suit their own imaginations, capabilities, and, importantly, even languages. This is perhaps what Chargaff sensed when, in 1987, he commented with respect to the "semi-industrial production of babies" that "the demand was less overwhelming than the desire on the part of the scientists to test their newly developed techniques" (Chargaff 1987, 199). He also noted, as Sydney Brenner did at Asilomar and many others have done since then, that scientific advances involving much money changing hands must also change the "ethical aroma" of science. Yet, one recurrent simplification that science, aided and abetted by the mass media, performs about itself is to rid its progress narrative of all material incentives and supports – whether the hands of the modest, invisible technicians who turn idea into product, or the spin-offs and start-ups that have turned the modern university into a hotbed of business entrepreneurship, or the research agendas funded by Silicon Valley's tech titans who want to buy knowledge in order to lengthen their own unimaginably privileged lifespans.

The life that emerges as the object of study from this tangled, instrumental, commercially driven technoscience is not the life that young Shvetaketu, fresh from his immersion in the Vedas, contemplated when he opened the fruit of the *nyagrodha* tree and found inside the nothing that was said to be the essence of all being. Nor is it the cycle of life, from origin to end, that Gauguin imagined when he painted his immortal Tahitian masterpiece. Once again, the point of these examples is not to privilege any specific rendition of life's meaning as intrinsically better or worse than others, or more or less true or worthy. It is to observe that those multiple readings exist for reasons we prize as human beings, and they rightfully configure our moral and political imaginations. Biology that claims a monopoly on defining life flattens an indescribably layered and complex concept when it decontextualizes what life *is* from eons of reflection on what life is *for*. Unless those ideas too are given privileged seats at the tables of collective sense-making, the lives we produce or save or prolong with scientific and technological know-how may not be the ones we deem most worth living.

At this point, the pragmatic and skeptical reader may well say that decontextualization is precisely what is called for in order to make progress in understanding life as a physical phenomenon, so people can gain the tools with which to live their lives more freely and knowingly, unencumbered by nature's manifold cruelties. After all, the rise in the production of IVF babies represents just such a market choice by people deprived of the rewards of natural childbirth; the technique has enabled millions of conceptions that could not otherwise have taken place and has greatly increased the sum total of human happiness. CRISPR, with its precise and efficient editorial power, now seems poised to perform even greater miracles of correcting nature's transcription errors, to the point of allowing us to eradicate pesky species. But these, we have seen, are partial stories about partial triumphs. How far we should go with human genome editing is not a question on which there is even scientific agreement, let alone on the harder questions that feminists and other critical theorists have raised of what counts as happiness, or parental fulfillment, or the ideal family and society.

With the rise of IVF came a series of dilemmas about how to manage production's side products: the shadowy hordes of frozen embryos, the cell lines derived from their bodies, and now the constructs of synthetic biology that could mimic embryonic life without ever having participated in life's natural cycles. There is nothing wrong with any of these developments in and of themselves, and many promise to alleviate genuine misery for some. But the contentious, unresolved quarrels that still swirl around these activities, and the not-so-private fears of cutting-edge biologists that their work could be shut down by ignorant populism, speak to persistent doubts within science, as well as to grave shortcomings in the relations between science and society. Those shortcomings are rooted in a linear idea of progress that too often lets the terms of debate be defined by what Stephen Toulmin once called science's "ecclesiastical courts" (see Culliton 1976, 451), instead of in messier consultation, and possibly long-drawn contestation, with democratic and dissident publics.

How to engage wider involvement in the governance of biological invention in this period of rapid technological advance remains a big puzzle for the human community. One idea that is gaining ground is the need to supplement existing institutional infrastructures, such as national high courts and bioethics bodies, to enable conversations that are both more inclusive and more respectful of divergent perspectives on the meaning and ends of life. A global observatory on genome editing and other heredity-altering techniques, for instance, might collect reflections from widely divergent legal and political cultures, analyze and disseminate perspectives from varied disciplinary standpoints, and facilitate a more cosmopolitan approach to ethical reflection on life's purposes (Jasanoff and Hurlbut 2018). Such a forum would help counter the reality that, even in our information-soaked age, we know very little of the multiple histories and cultures that have shaped human attempts to answer the questions Gauguin took with him from Paris to Tahiti: *Where Do We Come From / What Are We / Where Are We Going?*

Postlude

Most of us love life. We fear death, our children's and our friends' more than our own. We do not like to see misery in the lives of those already existing in the world – neither hunger nor disease nor mental illness nor premature death. The life that each of us holds as precious, however, is not the bare life of blood, bone, or even DNA. It is a life that is lived in relationships, intimate, familial, communal, national, and occasionally global.

Even Schrödinger, the physicist who did so much to turn biology away from organismic holism, ended his reflection on what life is by attending to its embedding in relationships, with one's remembered youth, with past and current friendships, with characters in the novels one reads. Life, especially but not only human life, is bound up too with ideas of a shared project, such as the dignity and integrity of the human and the sustainability of the Earth that supports all hitherto known forms of life. To appreciate life deeply has long been to make sense of what makes life valuable, not to oneself alone but in relation to others' ways of being.

We are, as well, creative and communicative beings, with the means and inclination to reflect on our own condition, often with no greater end in mind than to fulfill that ancient mandate to "know thyself." Often, the urge to know is coupled in a good world with making things that not only repair or prolong life, but also adorn and enrich our understanding of it: art, music, film, novels, essays, poetry, and philosophy. Often, too, collective reflection is needed to achieve well-ordered societies that enhance and protect the value of life, through the passage of laws and the design of safeguards against abuses of power. Lawmakers from Hammurabi to the present day understood and took on that responsibility. Without those ordering hands, we would all be outcasts, and *bios*, life in relation to the contexts that define its worth, would be an impossibility. It remains a universal condition of the good society that it must encourage the flourishing of life in all its varied meanings.

Science exerts power in part by turning the myriad pathways for living that humanity has evolved over millennia into singular channels that have undeniable value

Life's Purposes

for segments of the human community, such as eliminating inherited disease or producing more biofuels, but these "solutions" may not speak to the fundamentals of the human condition, and they may err or produce unintended consequences through premature simplification. Seduced by science's power to know and name its inventions, we are in danger of forgetting that the reaches of the human mind extend far beyond the electron microscope and the Lego bricks of synthetic biology. Even Borges's Library of Babel held only works composed in the twenty-six letters of the Roman alphabet. Raised in a language written in a different syllabary, but with a musicality of its own that rivals any other, I think of the meditations on life learned in even the most secular Bengali families of the twentieth century, where the materiality of life was treated as almost incidental to the spirit that animates living. I think in particular of *Aguner Poroshmoni* (Touchstone of Fire), the 1905 song by Rabindranath Tagore that accompanied (and perhaps still does) most memorial services at the passing of family or friends, with its opening stanza:

Purify my life
With the purging touch of fire
Purify my life
With your blessings of searing pain
Make it pure like the gold
That passes the test of fire.[1]

Notes

Chapter 1   A New Lens on Life

1 In French, with all capitals and no question marks: *D'où Venons Nous / Que Sommes Nous / Où Allons Nous*. The painting can be seen in Boston's Museum of Fine Arts.

2 The sixth and final edition published by Darwin in 1872 changed the title to *The Origin of Species*.

3 James Watson-Biographical, Nobelprize.org; https://www.nobelprize.org/nobel_prizes/medicine/laureates/1962/watson-bio.html (accessed August 2017).

4 Under the Bayh–Dole Act, states retained "march-in" rights to override privately held patents in extraordinary circumstances, such as the outbreak of an epidemic disease.

5 The White House, Office of the Press Secretary, "Remarks Made by the President, Prime Minister Tony Blair of England (via satellite), Dr. Francis Collins, Director of the National Human Genome Research Institute, and Dr. Craig Venter, President and Chief Scientific Officer, Celera Genomics Corporation, on the Completion of the First Survey of the Entire Human Genome Project," June 26, 2000; archived at https://www.genome.gov/10001356/june-2000-white-house-event/ (accessed September 2017).

6 https://www.buzzfeed.com/bfopinion/race-genetics-da vid-reich?utm_term=.ffA0vgD7K#.ftnoXm5bn (accessed June 2018).

Chapter 2   Book of Revelations

1 Modernized quote from Newton's letter to Robert Hooke of February 5, 1675; http://digitallibrary.hsp.org/ index.php/Detail/Object/Show/object_id/9285 (accessed September 2017).

2 September 28 is the date most frequently assigned to Fleming's discovery; it is the day he apparently picked himself, and it has been widely reproduced, although the American Chemical Society gives it as September 3. American Chemical Society, International Historic Chemical Landmarks, Discovery and Development of Penicillin, http://www.acs.org/content/acs/en/education/ whatischemistry/landmarks/flemingpenicillin.html (accessed September 2017).

3 https://www.sciencehistory.org/historical-profile/alex ander-fleming.

4 The quotation appears without further attribution in many sources. The National Library of Scotland posted the quote on its Facebook page on September 28, 2012, attributing it to "Alexander Fleming on his discovery of penicillin 84 years ago today"; https://www. facebook.com/NationalLibraryOfScotland/posts/38498 0048238135 (accessed September 2017).

5 In the world of science journalism, *The Double Helix* is seen as a revolutionary moment, almost as much as the discovery of the physical structure of DNA in biomedicine. Francis Crick wrote a furious letter to Watson, copied to the editor of Harvard University Press, the president of Harvard, and others, in which he charged, "Your view of history is that found in the lower class of women's maga-zines" (Weiner 2012). Harvard turned down the book

and Watson turned to a commercial press. The resulting publication, both pilloried and acclaimed, humanized the face of science and showed that science writing could be made appealing to a broad public.

6 Watson repeatedly referred to Wilkins as "slow" in his memoir, as in statements like these: "He appeared to enjoy slowly understating important arguments" and "Slowly and precisely he detailed how, in spite of much elaborate crystallographic analysis, little real progress had been made by Rosy ..." (Watson 1968, x, xx).

Chapter 3   Life and Law: Constitutional Turns
1 Law 196 states: "If a man put out the eye of another man, his eye shall be put out" (Avalon Project translation).
2 The Code of Hammurabi is often cited as an early source of the concept of insurance in providing that traders could, for an extra payment, be absolved of loss that occurred through no fault of their own.
3 Department of Health, Education and Welfare, National Institutes of Health, Recombinant DNA Research Guidelines, *Federal Register*, (41 131), 27915 (July 7, 1976).
4 *Foundation on Economic Trends v. Heckler*, 756 F.2d 143 (D.C. Cir. 1985), 160.
5 Center for Science, Technology, Medicine and Society, STEP Talk, April 2, 2013; http://cstms.berkeley.edu/curr ent-events/from-the-lab-to-the-field-history-and-regulat ion-of-biotechnological-applications-in-agriculture/ (accessed September 2017).
6 *Diamond v. Chakrabarty*, 447 U.S. 303 (1980), 309.
7 *Diamond v. Chakrabarty*, 447 U.S. 303 (1980), 317.
8 *President and Fellows of Harvard College v. Canada (Commissioner of Patents)* 2002 SCC 76.
9 *Harvard College v. Canada*, para. 163.
10 *Monsanto Canada Inc. v. Schmeiser* 2004 SCC 34.

11 *Association for Molecular Pathology et al. v. USPTO and Myriad Genetics*, 569 U.S. 576 (2013).

12 Justice Antonin Scalia filed a separate opinion, concurring in the judgment.

13 NIH director Francis Collins and deputy director Kathy Hudson made it clear that this settlement would not set a precedent for any future NIH-funded research with cell lines (Hudson and Collins 2013).

Chapter 4   Life in the Gray Zone

1 This technique is known as somatic cell nuclear transfer, because the nucleus from a somatic cell (that is, not a reproductive cell or gamete) is transferred into a different host cell.

2 At the time of this writing, the first British embryos of this kind were scheduled to be created in a Newcastle fertility research center in 2018, using eggs from women known to be carrying the DNA for a rare neurodegenerative disease called Merff syndrome (Sample 2018).

3 Exact numbers are hard to come by, since there is no global census of IVF children, but one estimate put the number at 6.5 million as of mid-2017. *Focus on Reproduction*, European Society of Human Reproduction and Embryology (ESHRE), July 5, 2016; https://focusonreproduction.eu/2016/07/05/6-5-million-ivf-babies-since-louise-brown/ (accessed February 2018).

4 Section 509A of Public Law 104–99, January 26, 1996

5 "President Discusses Stem Cell Research," The Bush Ranch, Crawford, Texas, August 9, 2001; https://georgewbush-whitehouse.archives.gov/news/releases/2001/08/20010809-2.html (accessed March 2018).

6 "Obama's Remarks on Stem Cell Research," The White House, *New York Times*, March 9, 2009; http://www.nytimes.com/2009/03/09/us/politics/09text-obama.html (accessed March 2018).

7 *Sherley v. Sebelius*, 704 F. Supp. 2d 63 (D.D.C. 2010), at p. 71.
8 English translation from https://www.btg-bestellservice. de/pdf/80201000.pdf (accessed February 2018).

Chapter 5   Language Games

1 Genesis 2:19 (KJV).
2 See https://www.bio.org/articles/history-biotechnology (accessed March 2018).
3 See http://www.biotechinstitute.org/go.cfm?do=Page.Vi ew&pid=22 (accessed March 2018).
4 See http://www.biotechnology.amgen.com/timeline.html and http://www.biotechnology.amgen.com/biotechnol ogy-explained.html (accessed March 2018).
5 Lord Kennet, Lords, January 15, 1988, col. 1497.
6 Human Fertilisation and Embryology Act 1990, Section 1.
7 The original manuscript still used the term SHELE, and that was the term used by the reviewers. In their response, the authors glossed over their terminological change in one quick sentence: "A common element of feedback we heard from both reviewers and others who have given us comments has been that our proposed processes made too tight a connection between the exploratory inquiries into the conceptual/moral issues raised by SHELEs (now SHEEFs) and the determination of guidelines for regulating SHEEF experiments." See Author Response, https:// elifesciences.org/articles/20674.

Chapter 6   A New Biopower

1 This was actually a second birth for biological engineering at MIT. A new research and teaching program in biological engineering, combining engineering with basic knowledge of physics, chemistry, and mathematics, was established in 1937 (Bud 1993, 86). Between

1942 and 1944, biological studies were conducted in the Department of Biology and Biological Engineering, but the department name then reverted back to Biology; https:// libraries.mit.edu/mithistory/research/schools-and-depart ments/school-of-science/department-of-biology/ (accessed February 2018).

2 "Engineering Biology," A Talk with Drew Endy, *Edge*, February 17, 2008; https://www.edge.org/conversation/ drew_endy-engineering-biology (accessed February 2018).

3 The description appears on Synberc's welcome page; https://www.synberc.org/.

4 Kevin Esvelt, "When Are We Obligated To Edit Wild Creatures?" leapsmag, May 28, 2018; https://leapsmag. com/when-are-we-obligated-to-edit-wild-creatures/ (accessed June 2018).

5 The White House, National Bioeconomy Blueprint, April 2012, p. 7; https://obamawhitehouse.archives.gov/sites/ default/files/microsites/ostp/national_bioeconomy_bluepr int_april_2012.pdf (accessed June 2018).

6 Carl Woese was not a member of the NRC committee that produced the New Biology report. The report's co-chairs were Thomas Connelly of DuPont Company and MIT biologist and Nobel laureate Phillip Sharp.

7 The origin of the term "molecular biology" is often traced to a 1938 report by the Rockefeller Foundation's Warren Weaver (1970).

8 Directive to reduce indirect land use change for biofuels and bioliquids ((EU)2015/1513); http://eur-lex.europa.eu/ legal-content/EN/TXT/?uri=CELEX%3A32015L1513 (accessed March 2018).

9 The Council is a private organization funded by both state and private resources, but its opinions carry considerable weight in a political system that has devolved bioethical thinking away from the centers of official power.

Chapter 7 Life's Purposes

1 This translation is by Anandamayee Majumdar and Rumela Sengupta, *Gitabitan in English*; http://gitabitan-en.blogspot.com/2011/09/philosophers-stone-of-fire.html (accessed August 2018).

References

Aach, John, Jeantine Lunshof, Eswar Iyer, and George M. Church. 2017. "Addressing the Ethical Issues Raised by Synthetic Human Entities with Embryo-like Features." eLife, March 21. https://elifesciences.org/articles/20674 (accessed March 2018).

ACS (American Chemical Society) and RSC (Royal Society of Chemistry). 1999. "The Discovery and Development of Penicillin 1928–1945." https://www.acs.org/content/dam/acsorg/education/whatischemistry/landmarks/flemingpenicillin/the-discovery-and-development-of-penicillin-commemo rative-booklet.pdf (accessed September 2018).

Allen, Garland E. 1999. "Modern Biological Determinism: The Violence Initiative." In Michael Fortun and Everett Mendelsohn, eds., The Practices of Human Genetics. Dordrecht: Kluwer, pp. 1–23.

Anderson, Benedict. 1983. Imagined Communities. London: Verso.

Arendt, Hannah. 1958. The Human Condition. Chicago, IL: University of Chicago Press.

References

Associated Press. 1987. "Altered Bacteria Fight Frost." *New York Times*, June 10.

Basile, Jonathan. 2015. "The Library of Babel as Seen from Within." *Paris Review*, July 23.

Berg, Paul. 2008. "Meetings that Changed the World: Asilomar 1975: DNA Modification Secured." *Nature* 455: 290–291.

Berg, Paul, David Baltimore, Sydney Brenner, Richard O. Roblin, and Maxine F. Singer. 1975. "Summary Statement of the Asilomar Conference on Recombinant DNA Molecules." *Proceedings of the National Academy of Sciences* 72(6): 1981–1984.

Borges, Jorge Luis. 1962. *Labyrinths: Selected Stories and Other Writings*, ed. Donald A. Yates and James E. Irby. New York: New Directions.

Boseley, Sarah. 2008. "Balancing Act." *Guardian*, May 28.

Bud, Robert. 1993. *The Uses of Life: A History of Biotechnology*. Cambridge: Cambridge University Press.

Bush, Vannevar. 1945. *Science: The Endless Frontier*. Washington, DC: US Government Printing Office.

Carson, Rachel. 1962. *Silent Spring*. Boston, MA: Houghton Mifflin.

CDC (Centers for Disease Control and Prevention). 2018. "Suicide Rising Across the US: More than a Mental Health Concern." https://www.cdc.gov/vital signs/suicide/index.html (accessed June 2018).

Chargaff, Erwin. 1976. "On the Dangers of Genetic Meddling." *Science* 192(4243): 938–940.

Chargaff, Erwin. 1987. "Engineering a Molecular Nightmare." *Nature* 327: 199–200.

Chichilnisky, Graciela, and Geoffrey Heal. 1998.

"Economic Returns from the Biosphere." *Nature* 391: 629–630.

Chong, Curtis. 1995. "Professor Finds Beauty in Violins and Viruses." *Harvard Crimson*, November 22.

Cole, Robert J. 1980. "Genentech, New Issue, Up Sharply." *New York Times*, October 15.

Conway, Gordon. 1999. *The Doubly Green Revolution*. Ithaca, NY: Cornell University Press.

Costanza, Robert, Ralph d'Arge, Rudolf de Groot, Stephen Farber, Monica Grasso, ... Marjan van den Belt. 1997. "The Value of the World's Ecosystem Services and Natural Capital." *Nature* 387: 253–260.

Crick, Francis. 1974. "The Double Helix: A Personal View." *Nature* 248: 766–769.

Culliton, Barbara J. 1976. "Public Participation in Science: Still in Need of Definition." *Science* 192(4238): 451–453.

Daniels, Mitch. 2017. "Avoiding GMOs Isn't Just Anti-Science. It's Immoral." *Washington Post*, December 27.

Darwin, Charles. 1859. *On the Origin of Species by Means of Natural Selection, or the Preservation of Favoured Races in the Struggle for Life*. London: John Murray.

Darwin, Charles. 1860. *A Naturalist's Voyage Round the World*. London: John Murray.

Dennett, Daniel C. 1995. *Darwin's Dangerous Idea: Evolution and the Meanings of Life*. New York: Simon and Schuster.

Doudna, Jennifer A., and Samuel H. Sternberg. 2017. *A Crack in Creation: Gene Editing and the Unthinkable Power to Control Evolution*. New York: Houghton Mifflin.

References

ESHRE Guideline Group on Good Practice in IVF Labs. 2015. *Revised Guidelines for Good Practice in IVF Laboratories (2015)*. Grimbergen, Belgium: ESHRE.

Espeland, Wendy N. 1998. *The Struggle for Water: Politics, Rationality, and Identity in the American Southwest*. Chicago, IL: University of Chicago Press.

Fleck, Ludwik. 1981 [1935]. *Genesis and Development of a Scientific Fact*. Chicago, IL: University of Chicago Press.

Foucault, Michel. 1969 (trans. A. M. Sheridan Smith, 1972). *Archaeology of Knowledge and the Discourse on Language*. New York: Pantheon.

Foucault, Michel. 1990 [1976]. *The History of Sexuality*, vol. 1. New York: Vintage.

Fredrickson, Donald S. 1979. "A History of Recombinant DNA Guidelines in the United States." *Recombinant DNA Technical Bulletin* 2(2): 151–156. https://profiles.nlm.nih.gov/ff/b/b/k/c/_/ffbbkc.pdf (accessed September 2017).

Fredrickson, Donald S. 1991. "Asilomar and Recombinant DNA: The End of the Beginning." In Kathi E. Hanna, ed., *Biomedical Politics*. Washington, DC: National Academies Press, pp. 258–324.

Fredrickson, Donald S. 2001. *The Recombinant DNA Controversy: A Memoir: Science, Politics, and the Public Interest 1974–1981*. Washington, DC: American Society of Microbiology Press.

Freud, Sigmund. 1920. *A General Introduction to Psychoanalysis*, trans. G. Stanley Hall. New York: Boni and Liveright.

Friend, Tad. 2018. "Silicon Valley's Quest to Live Forever." *New Yorker*, April 3.

Gauguin, Paul, and Peter Russell. 2016. *Masters of*

References

Art: Paul Gauguin. Hastings, East Sussex: Delphi Classics.

Geertz, Clifford. 1975. "Common Sense as a Cultural System." *Antioch Review* 33(1): 5–26.

Gibbons, Michael, Camille Limoges, Helga Nowotny, Simon Schwartzman, Peter Scott, and Martin Trow. 1994. *The New Production of Knowledge.* London: Sage Publications.

Gillis, Justin. 2009. "Norman Borlaug, Plant Scientist Who Fought Famine, Dies at 95." *New York Times*, September 13. http://www.nytimes.com/2009/09/14/business/energy-environment/14borlaug.html.

Gitschier, Jane. 2009. "Wonderful Life: An Interview with Herb Boyer." *PLoS Genetics* 5(9). https://www.ncbi.nlm.nih.gov/pmc/articles/PMC2741595/ (accessed September 2017).

Graham, Loren. 2016. *Lysenko's Ghost: Epigenetics and Russia.* Cambridge, MA: Harvard University Press.

Greene, Jeremy A., and Kevin R. Riggs. 2015. "Why Is There No Generic Insulin? Historical Origins of a Modern Problem." *New England Journal of Medicine* 372: 1171–1175.

Hacking, Ian. 2006. "Making Up People." *London Review of Books* 28(16): 23–26.

Hilgartner, Stephen. 2017. *Reordering Life: Knowledge and Control in the Genomics Revolution.* Cambridge, MA: MIT Press.

Hood, Elizabeth E. 2016. "Plant-Based Biofuels." *F1000 Research.* https://f1000research.com/articles/5-185/v1 (accessed September 2018).

Horn, Dara. 2018. "The Men Who Want to Live Forever." *New York Times*, January 25. https://www.

References

nytimes.com/2018/01/25/opinion/sunday/silicon-vall ey-immortality.html (accessed September 2018).

Hudson, Kathy L., and Francis S. Collins. 2013. "Biospecimen Policy: Family Matters." *Nature* 500: 141–142.

Hunt-Grubbe, Charlotte. 2007. "The Elementary DNA of Dr. Watson." *Sunday Times*, October 14.

Hurlbut, J. Benjamin. 2015. "Remembering the Future: Science, Law and the Legacy of Asilomar." In Sheila Jasanoff and Sang-Hyun Kim, eds., *Dreamscapes of Modernity: Sociotechnical Imaginaries and the Fabrication of Power*. Chicago, IL: University of Chicago Press, pp. 126–151.

Hurlbut, J. Benjamin. 2017. *Experiments in Democracy: Human Embryo Research and the Politics of Bioethics*. New York: Columbia University Press.

Hurlbut, J. Benjamin, Insoo Hyun, Aaron D. Levine, Robin Lovell-Badge, Jeantine E. Lunshof, ... Laurie Zoloth. 2017. "Revisiting the Warnock Rule." *Nature Biotechnology* 35(11): 1029–1042.

Huxley, Aldous. 1932. *Brave New World*. London: Chatto and Windus.

Hyman, Steven. 2014. "Mental Health: Depression Needs Large Human-Genetics Studies." *Nature* 515: 189–191.

Jasanoff, Sheila. 1995a. "Product, Process, or Programme: Three Cultures and the Regulation of Biotechnology." In Martin Bauer, ed., *Resistance to New Technology*. Cambridge: Cambridge University Press, pp. 311–331.

Jasanoff, Sheila. 1995b. *Science at the Bar: Law, Science and Technology in America*. Cambridge, MA: Harvard University Press.

References

Jasanoff, Sheila. 2005. *Designs on Nature: Science and Democracy in Europe and the United States.* Princeton, NJ: Princeton University Press.

Jasanoff, Sheila. 2006 "Biotechnology and Empire: The Global Power of Seeds and Science." *Osiris* 21(1): 273–292.

Jasanoff, Sheila. 2016. *The Ethics of Invention.* New York: Norton.

Jasanoff, Sheila, and J. Benjamin Hurlbut. 2018. "A Global Observatory for Gene Editing." *Nature* 555: 435–437.

Jasanoff, Sheila, and Sang-Hyun Kim, eds. 2015. *Dreamscapes of Modernity: Sociotechnical Imaginaries and the Fabrication of Power.* Chicago, IL: University of Chicago Press.

Jasanoff, Sheila, and Ingrid Metzler. 2018. "Borderlands of Life: IVF Embryos and the Law in the United States, United Kingdom, and Germany." *Science, Technology and Human Values.* http://journals.sage pub.com/doi/10.1177/0162243917753990.

Jasanoff, Sheila, J. Benjamin Hurlbut, and Krishanu Saha. 2015. "CRISPR Democracy: Gene Editing and the Need for Inclusive Deliberation." *Issues in Science and Technology* XXXII(1): 25–32.

Judson, Horace F. 1975a. "Fearful of Science: After Copernicus, after Darwin, after Freud, Comes Molecular Biology. Is Nothing Sacred?" *Harper's Magazine* (March): 32–41.

Judson, Horace F. 1975b. "Fearful of Science: Who Shall Watch the Scientists?" *Harper's Magazine* (June): 70–76.

Judson, Horace F. 1979. *The Eighth Day of Creation: Makers of the Revolution in Biology.* New York:

Simon and Schuster.

Juma, Calestous. 2016. *Innovation and Its Enemies: Why People Resist New Technologies*. New York: Oxford University Press.

Kay, Lily E. 1993. *The Molecular Vision of Life: Caltech, the Rockefeller Foundation, and the Rise of the New Biology*. Oxford: Oxford University Press.

Kay, Lily E. 2000. *Who Wrote the Book of Life? A History of the Genetic Code*. Stanford, CA: Stanford University Press.

Keller, Evelyn Fox. 1985. *Reflections on Gender in Science*. New Haven, CT: Yale University Press.

Keller, Evelyn Fox. 2002. *Making Sense of Life: Explaining Biological Development with Models, Metaphors, and Machines*. Cambridge, MA: Harvard University Press.

Kemp, Martin. 2003. "The Mona Lisa of Modern Science." *Nature* 421(6921): 416–420.

Kohler, Robert E. 2002. *Landscapes and Labscapes: Exploring the Lab–Field Border in Biology*. Chicago, IL: University of Chicago Press.

Lahiri, Nayanjot. 2015. *Ashoka in Ancient India*. Cambridge, MA: Harvard University Press.

Landecker, Hannah. 2013. "The Life of Movement: From Microcinematography to Live-Cell Imaging." *Journal of Visual Culture* 11(3): 378–399.

Lax, Eric. 2004. *The Mold in Dr. Florey's Coat: The Story of the Penicillin Miracle*. New York: Henry Holt.

Lippman, Abby. 1991. "Prenatal Genetic Testing and Screening: Constructing Needs and Reinforcing Inequities." *American Journal of Law and Medicine* 17(1–2): 15–50.

References

MA (Millennium Ecosystem Assessment). 2005. *Ecosystems and Human Well-Being: Synthesis*. Washington, DC: Island Press.

Maasakkers, Mattijs van. 2016. *The Creation of Markets for Ecosystem Services in the United States: The Challenge of Trading Places*. London: Anthem Press.

Malins, Chris. 2017. *Thought for Food: A Review of the Interaction between Biofuel Consumption and Food Markets*. London: Cerulogy.

McElheny, Victor K. 2003. *Watson and DNA: Making a Scientific Revolution*. New York: Perseus Publishing.

McLean, Phillip. 1997. "The Recombinant DNA Debate – Historical Events in the rDNA Debate." https://www.ndsu.edu/pubweb/~mcclean/plsc431/debate/debate3.htm (accessed September 2017).

Merton, Robert K. 1973. "The Normative Structure of Science." In R. K. Merton, *The Sociology of Science: Theoretical and Empirical Investigations*. Chicago, IL: University of Chicago Press, pp. 267–278.

Miller, Claire Cain and Andrew Pollack. 2013. "Tech Titans Form Biotechnology Company." *New York Times*, September 18.

Miller, Henry I., and Frank E. Young. 1987. "Biotechnology: A 'Scientific' Term in Name Only." *Wall Street Journal*, January 13.

Min, Jason K., Paul Claman, and Ed Hughes. 2006. "Guidelines for the Number of Embryos to Transfer Following In Vitro Fertilization." *Journal of Obstetrics and Gynaecology Canada* 28(9): 799–813.

Morris, Meaghan, and Paul Patton, eds. 1979. *Michel Foucault: Power, Truth, Strategy*. Sydney: Feral.

Mulkay, Michael. 1997. *The Embryo Research Debate:*

Science and the Politics of Reproduction. Cambridge: Cambridge University Press.

Mullis, Kary. 1998. *Dancing Naked in the Mind Field.* New York: Pantheon.

NASEM (National Academies of Sciences, Engineering, and Medicine). 2017. *Human Genome Editing: Science, Ethics, and Governance.* Washington, DC: The National Academies Press. doi: https://doi.org/10.17226/24623.

NBAC (National Bioethics Advisory Commission). 1997. *Cloning Human Beings.* Rockville, MD: NBAC.

NRC (National Research Council). 2009. *A New Biology for the 21st Century.* Washington, DC: National Academies Press.

Nuffield Council on Bioethics. 2012. *Emerging Biotechnologies: Technology, Choice and the Public Good.* London: Nuffield Council on Bioethics.

Paarlberg, Robert. 2008. *Starved for Science: How Biotechnology Is Being Kept Out of Africa.* Cambridge, MA: Harvard University Press.

PCSBI (Presidential Commission for the Study of Bioethical Issues). 2010. *New Directions: The Ethics of Synthetic Biology and Emerging Technologies.* Washington, DC: PCSBI.

Polanyi, Michael. 1962. "The Republic of Science." *Minerva* 1: 54–73.

Powell, Alvin. 2016. "Updating Embryo Research Guidelines." *Harvard Gazette*, November 18. https://news.harvard.edu/gazette/story/2016/11/updating-embryo-research-guidelines/ (accessed March 2018).

President's Commission for the Study of Ethical Problems in Medicine and Biomedical and Behavioral

Research. 1982. *Splicing Life: A Report on the Social and Ethical Issues of Genetic Engineering with Human Beings.* Washington, DC: US Government Printing Office.

Rabinow, Paul. 1996. *Making PCR: A Story of Biotechnology.* Chicago, IL: University of Chicago Press.

Reardon, Jenny. 2004. *Race to the Finish: Identity and Governance in an Age of Genomics.* Princeton, NJ: Princeton University Press.

Regalado, Antonio. 2017. "Artificial Human Embryos Are Coming, and No One Knows How to Handle Them." *Technology Review*, September 19.

Reich, David. 2018a. *Who We Are and How We Got Here: Ancient DNA and the New Science of the Human Past.* New York: Pantheon.

Reich, David. 2018b. "How Genetics Is Changing Our Understanding of 'Race'." *New York Times*, March 23.

Rheinberger, Hans-Jörg. 1997. *Toward a History of Epistemic Things: Synthesizing Proteins in the Test Tube.* Stanford, CA: Stanford University Press.

Richards, Robert J. 1983. "Why Darwin Delayed, or Interesting Problems and Models in the History of Science." *Journal of the History of the Behavioral Sciences* 19: 45–53.

Richardson, John. 2009. "Gauguin's Last Testament." *Vanity Fair*, May 18.

Roberts, Leslie. 1992. "Why Watson Quit as Project Head." *Science* 256(5005): 301–302.

Roosth, Sophia. 2017. *Synthetic: How Life Got Made.* Chicago, IL: University of Chicago Press.

Rossiter, Margaret W. 1984. *Women Scientists in*

America: Struggles and Strategies to 1940. Baltimore, MD: Johns Hopkins University Press.

Ruse, Michael. 1981. *Is Science Sexist? And Other Problems in the Biomedical Sciences.* Dordrecht, NL: Reidel.

Sagoff, Mark. 2005. "The Catskills Parable." PERC Report 23(2). https://www.perc.org/2005/06/01/the-catskills-parable/ (accessed September 2018).

Sample, Ian. 2018. "UK Doctors Select First Women to Have 'Three-Person Babies'." *Guardian*, February 1.

Sarewitz, Daniel. 2016. "Saving Science." *New Atlantis* (Spring/Summer 2016): 5–40.

Schrödinger, Erwin. 1967 [1944]. *What Is Life?* Cambridge: Cambridge University Press.

Sen, Amartya. 1981. *Poverty and Famines: An Essay on Entitlement and Deprivation.* Oxford: Oxford University Press.

Shapin, Steven, 1999. "Nobel Savage." *London Review of Books* 21(13): 17–18.

Shapin, Steven and Simon Schaffer. 1985. *Leviathan and the Air-Pump: Hobbes, Boyle, and the Experimental Life.* Princeton, NJ: Princeton University Press.

Shelley, Mary Wollstonecraft. 1998 [1818]. *Frankenstein, Or, The Modern Prometheus: The 1818 Text.* Oxford: Oxford University Press.

Shiva, Vandana. 1991. *The Violence of the Green Revolution: Third World Agriculture, Ecology, and Politics.* Penang, Malaysia: Third World Network.

Singer, Maxine S. 1984. "Genetics and the Law: A Scientist's View." *Yale Law and Policy Review* 3(2): 315–335.

Sinsheimer. Robert L. 1976. "Recombinant DNA: On Our Own." *BioScience* 26(10): 599.

References

Sinsheimer, Robert L. 1994. *The Strands of a Life.* Berkeley: University of California Press.

Skloot, Rebecca. 2010. *The Immortal Life of Henrietta Lacks.* New York: Crown.

Specter, Michael. 2014. "Seeds of Doubt." *New Yorker,* August 24, p. 46.

Stein, Rob. 2017. "Embryo Experiments Reveal Earliest Human Development, but Stir Ethical Debate." *National Public Radio,* March 2. https://www.npr.org/sections/health-shots/2017/03/02/516280895/embryo-experiments-reveal-earliest-human-development-but-stir-ethical-debate (accessed March 2018).

Steinfels, Peter. 1976. "Biomedical Research and the Public: A Report from the Airlie House Conference." *Hastings Center Report* 6(3): 21–25.

Strasser, Bruno J. 2003. "Who Cares About the Double Helix?" *Nature* 422: 803–804.

Sulloway, Frank J. 1982. "Darwin and His Finches: The Evolution of a Legend." *Journal of the History of Biology* 15(1): 1–5.

Taylor, Charles. 2004. *Modern Social Imaginaries.* Durham, NC: Duke University Press.

Thapar, Romila. 1997. *Asoka and the Decline of the Mauryas.* Delhi: Oxford University Press.

Traub, James. 2005. "Lawrence Summers, Provocateur." *New York Times,* January 23.

Wade, Nicholas. 1973. "Microbiology: Hazardous Profession Faces New Uncertainties," *Science* 182(4112): 566–567.

Wade, Nicholas. 1975. "Genetics: Conference Sets Strict Controls to Replace Moratorium." *Science* 187(4180): 931–935.

Wade, Nicholas. 1998. "Scientist at Work/Kary Mullis:

After the 'Eureka,' a Nobelist Drops Out." *New York Times*, September 15.

Wade, Nicholas. 2000a. "Reading the Book of Life: The Overview; Genetic Code of Human Life Is Cracked by Scientists." *New York Times*, June 27.

Wade, Nicholas. 2000b. "Reading the Book of Life: A Historic Quest; Double Landmarks for Watson: Helix and Genome." *New York Times*, June 27.

Walters, Donna K. H. 1986. "Biotech Firms Struggle With Identity Crisis: Definition of Industry Affects How Products Are Regulated by Government." *Los Angeles Times*, October 27. http://articles.latimes.com/1986-10-27/business/fi-7566_1_biotechnology-companies (accessed March 2018).

Warnock, Mary. 1984. *The Report of the Committee of Inquiry into Human Fertilisation and Embryology*. London: Her Majesty's Stationery Office.

Watson, James D. 1968. *The Double Helix: A Personal Account of the Discovery of the Structure of DNA*. New York: Atheneum.

Watson, James D., and Francis H. C. Crick. 1953. "A Structure for Deoxyribose Nucleic Acid." *Nature* 171(4356): 737–738.

Weaver, Warren. 1970. "Molecular Biology: Origin of the Term." *Science* 170(3958): 581–582.

Weiner, Jonathan. 2012. "Laboratory Confidential." *Columbia Journalism Review*. http://archives.cjr.org/second_read/laboratory_confidential.php (accessed September 2018).

Weintraub, Karen. 2017. "Ethical Guidelines on Lab-Grown Embryos Beg for Revamping, Scientists Say." *Scientific American*, March 21.

Whaley, Lauren M. 2017. "The Emptiness of the All-

Male Panel." *Undark*, June 6, https://undark.org/ article/manels-all-male-panel/ (accessed March 2018).

Wilson, A. N. 2017. *Charles Darwin: Victorian Mythmaker*. London: Hodder and Stoughton.

Winickoff, David, Sheila Jasanoff, Lawrence Busch, Robin Grove-White, and Brian Wynne. 2005. "Adjudicating the GM Food Wars: Science, Risk, and Democracy in World Trade Law." *Yale Journal of International Law* 30: 81–123.

Wittgenstein, Ludwig. 1953. *Philosophical Investigations*, trans. G. E. M. Anscombe. Oxford: Blackwell.

Woese, Carl R. 2004. "A New Biology for a New Century." *Microbiology and Molecular Biology Reviews* 68(2): 173–186.

Wright, Susan. 1994. *Molecular Politics: Developing American and British Regulatory Policy for Genetic Engineering, 1972–1982*. Chicago, IL: University of Chicago Press.

Yoffe, Emily. 1994. "Is Kary Mullis God (or Just the Big Kahuna?)" *Esquire*, July 1994, p. 68.

Zimmer, Carl. 2017. "A New Form of Stem-Cell Engineering Raises Ethical Questions." *New York Times*, March 21.

Ziolkowski, Theodore. 1997. *The Mirror of Justice: Literary Reflections of Legal Crises*. Princeton, NJ: Princeton University Press.

Index

Index

Index

Index

Index

Index

Index